Women in Leadership

Essential skills to become a great femaleand the workplace, Lead successful teams, and inspire employees

Romuald Fisher

CONTENTS

Women in LeadershiP

Introduction

Chapter One: Overview of Leadership

Chapter Two: The Difference Between Managing and Leading People

Chapter Three: Managing Difficult People and Situations to Succeed

Chapter Four: Overcoming Self-doubt and Proven Techniques

Chapter Five: How to Build High-performing Teams Using Coaching and Development

Chapter Six: Build Confidence in Yourself and Help Your Team Become a High-performing Team

Chapter Seven: Effective Communication Skills and Most Proven Techniques

Chapter Eight: Negotiation Tactics and Dealing with Difficult People for a Successful Outcome

Chapter Nine: Change Management and Overcoming Fear

Conclusion

INTRODUCTION

"If you don't take risks, you will never leave your comfort zone, and if you've never experienced failure, then it means you have never tried anything new. How could life be worth living without any expectations or the excitement of waiting to experience something new? I think you'll find that you are stronger than what you think, and you're capable of anything if you set your mind to it. NEVER LET ANYONE TELL YOU OTHERWISE!"

Those were the words my first manager Annie Blake spoke to me ten years ago. Everything she told me that day made it feel like something was waking up inside of me, especially the last line, which flipped on the switch for me.

I've learned that once you are presented with an opportunity, what you do with it matters. You are the only one who has control over your life; you decide what success means to you and define who you become.

As a young girl growing up in a family with more boys, I've always envied how the boys could choose to be anything they wanted to be. As for me, I've always been passionate about sports and fitness but was made to believe it looked better on men, and I am not strong enough for that. Alas, my dream would never come true simply because I was a woman. If only I knew that would take a toll on me and affect my career.

The bias was obvious, even if it wasn't intended. The worst part of it all was my life began to be tailored for me with a design that wasn't mine. It continued until I started working at a firm with more men in key positions. I knew I had the skills to seek promotion and work on important projects, but I was always scared of being too forward or speaking up.

Mrs. Annie was a manager I always admired in the company. I've always wished I could be half as confident as she was. We exchanged pleasantries occasionally, and I didn't realize she noticed I was dimming my light and sinking into where I shouldn't be. Well, I guess I was somewhat important. She managed to get a confession out of me after several persuasions. I remember crying out my heart while explaining how I felt to her and the struggles I face. I've yet to ask her how she understood all the gibberish I said while mixing words and snorts.

After comforting me, she left me with that strong message that has forever changed my life. Those words sent me out on a mission to become all I've ever wanted to be.

Right now, you may be going through a tough time coping with the idea of men being promoted at work and given the important tasks, enjoying bonuses and raises while you don't get to enjoy any of these when you know you have the skills and can produce the same results or even more if given a chance.

Perhaps you find it challenging to express yourself or not speak up at meetings for fear of not offending the top executives. If any of these scenarios sound anything like you, you are in luck.

You are reading the perfect book that will completely change your mindset and overhaul your life!

Today, we see an incredibly male-dominated world, which makes it important now more than ever to improve on the needed skills to become effective leaders and start getting all we deserve.

Women are gradually becoming a dominant force in the workplace despite the challenges. We are rising to key management positions in the public and private sectors. In fact, some have resolved to break the traditional ceiling, which has prevented them from entering those top leadership positions even when they have the requisite talents and skills to be there.

As leaders, we drive our teams to achieve and exceed our targets. Doing this requires innovation, creativity, professionalism, and the ability to get the best from our teams. When in leadership positions, we provide structural and cultural differences, skills, and imaginative perspectives, which help drive effective solutions.

This is evident in the Fortune 500, where women CEOs make up 6.4 percent of the list in America. In addition, we've realized that companies with a higher representation of women on their boards outperform organizations that don't. In fact, several studies have revealed that companies with higher gender diversity in their workforce and among senior leaders have more significant profits than those without.

There have been calls for more women to occupy leadership positions to close the gender wage gap for many years now. Only a few women occupy leadership roles, which doesn't sit well with me. I understand that organizations should create better policies and opportunities for women. Still, as women, we need to step forward and overcome those things holding us back.

We can see how women are now evolving and reaching big milestones across various industries these days. We've witnessed the advent of great women leaders such as Kamala Harris, Oprah Winfrey, Hillary Rodham Clinton, Christine Lagarde, and Theresa May, among others.

Even if most of us won't become international stars, we can break ground in our various industries, make a mark, and take the mantle of leadership without fear.

Yes, you can! You need easy and proven skills and traits that will help improve your self-confidence, sharpen your communication and boundary-setting skills, and learn how to effectively manage low-performing teams.

This book will improve your leadership skills needed to succeed in today's challenging world. It will be offering you fine-tuned strategies for effective leadership that have proven to work in business. We'll emphasize producing high-performing teams that positively affect business and organization. At the end of this book, you should've mastered how to overcome your fears and doubts, learn how to communicate effectively, and turn your emotions into successful actions.

Also, we'll explore and leverage modern leadership tools that will help your organization and businesses achieve high performance. We'll cover management concepts, essentials of effective leadership, and creative problem-solving skills that help in tactful negotiation.

You may be wondering why this subject matters to me.

The truth is, I've always been highly motivated about career development and leading people. Over the years, I've met people who positively and negatively impacted me. While on this journey, I got inspired by leaders who encouraged me and made me realize that I could tackle anything I wanted to. Sadly, they had more confidence in me than I had in myself then. Still, I am glad that the lessons I learned during those times resonated with me, and I've been abiding by the guidance I've received until today.

I know that my decision to become a leader can affect other aspects of my life. It can affect my role as a mother, partner, or spouse and my position in my family. It's like accepting a daily challenge to constantly strive to be a better person and leader.

I've learned that as a woman, I don't need to be afraid of becoming a leader and don't expect to be treated differently because of my gender. I am no different from others! I've dared to be extraordinary, no matter how bumpy the road is. I believe in myself and move with conviction. I never underestimate my ability to achieve great things, which has worked for me over the years.

I am writing this book because I've lived through career setbacks and conquered them to live a better life. I have experienced a better way of living and want you to do the same. I understand different challenges throughout life with managing people, and I want to teach others from my failures.

Many women struggle to be recognized; they want their voices to be heard, build an alliance, and overcome imposter syndrome. I want to motivate and fill a need by helping my target audience identify and understand the challenges they face, push through those barriers, serve their mission, and reach their full potential.

Leadership isn't a cakewalk. On the contrary, it is one of the most challenging tasks I've had to deal with. Nevertheless, I was able to extract the best out of my team and be in charge during difficult times. No doubt, we all want to be leaders, but have you considered the challenges you may be facing in that role? First, we need to understand that a leader is a captain that anchors a ship in the right direction; a wrong decision can sink the entire ship, and the entire blame is on you. This makes it crucial to understand what leadership entails and how to lead while overcoming obstacles.

Are you ready to be the face of your team by ensuring that your subordinates perform their duties diligently and are happy with their work? Whether you are already in a leadership role or simply want to elevate your career to new heights, we are about to start a journey that will address the challenges you may face and help you ignite your impact as a woman in your organization.

Let's get started!

CHAPTER ONE: OVERVIEW OF LEADERSHIP

"Leadership is about making others better as a result of your presence and making sure that impact lasts in your absence."

– Sheryl Sandberg

Every organization needs a leader regardless of its function and size. An organization without a leader is just like a *"muddle of men and machine,"* a society without a leader is a dangerous place to live in, and a country without leadership is simply anarchy.

Strong leadership is important for the growth of any organization. While leadership theories and principles haven't changed, issues women leaders face have also remained constant. We face realities such as pay gaps, misogyny, family expectations and pressures, uneven opportunities for growth, and cultural norms, which have created extra challenges.

As women, we are powerful agents of change that have been greatly underrepresented in decision-making concerning communities, businesses, and politics. Despite the challenges we face, our leadership has proven to be significantly instrumental to the success of an organization. Several studies have shown that encouraging diversity in the workplace brings different perspectives to an organization, which as a result, encourages innovation, boosts creativity, and inspires new ways of handling things. This leads to effective decision-making and overall success.

It is also worth mentioning that women have proven to outscore men on leadership competencies during a crisis. Therefore, it only makes sense for organizations to encourage women's leadership by putting up the policies, systems, and structures to ensure diversity in the workforce.

In this chapter, we will be discussing what leadership entails, what it means, the types, styles, and importance of female leadership, and the traits of successful female leaders. To be like something, you have to understand it and know it. Therefore, knowing these traits is the first step to being successful. But before we get to that, let's quickly look at what leadership and what being a leader entails.

What Is Leadership?

According to Keith Davis, *"Leadership is the ability to persuade others to seek defined objectives enthusiastically. It is the human factor which binds a group together and motivates it towards goals."*

Leadership is the process where a manager/executive/leader guides, directs, and influences the behavior and work of others (followers/subordinates) to achieve a specific goal. It is also the act of inducing subordinates to work with the zeal and confidence to drive results.

Leadership can also be defined as having the capacity to influence certain people toward achieving a collective goal. Leaders create future visions and then motivate their team members to achieve the vision created. The role of leadership shouldn't be gender-specific. It should be the set of leadership qualities an individual has cultivated and how they've developed themselves into a great leader that should be considered.

A leader is an individual who encourages and influences a group or team to realize a specific goal. Just as Chester Barnard summed it up, *"Leadership is the ability of a superior to influence the behavior of subordinates or groups and persuade them to follow a particular course of action."*

Next, we will be discussing leadership styles and why you need to identify yours.

Defining Leadership Styles

If I may ask, what is your idea of a good leader?

Your answer to this question will reveal a lot about your leadership style.

Leadership should be a fluid practice where there is a continuous change in how you help your organization grow. The longer you assume a leadership role, the more likely you'll change how you answer the question I just posed.

To become a successful leader, you need to know your standpoint. Knowing different leadership styles will help you figure out what type of leader you are.

A leadership style is your behavior and method when motivating, directing, and managing others. Your leadership style determines how you plan, strategize, and implement what you've planned while considering your team's well-being.

Whether you are leading a team, project, meeting, or organization, you need to identify your leadership style. Knowing different leadership styles makes it easy to know the one that suits you best. Take your time to familiarize yourself with them, and you will know the areas you need to improve on.

In fact, you may get to identify other ways you can lead to better serve your goals and better understand how to work with other leaders with a different leadership style.

Types of Leadership Styles

We are all unique, which makes every leader's approach to leadership unique. Your leadership style is a result of your personality. While some leaders are lenient, others are strict. While some are high-strung, others are mellow. The IMD has suggested that leadership styles be categorized according to personality traits.

Many leaders develop their leadership style based on their personality, experience, the unique needs of their organization, and their organizational culture. While leaders are different, the following are common leadership styles used in the workplace.

1. Democratic Leadership

This is an effective leadership style that is exactly as the name sounds. Here, the leader makes a decision based on the opinion of each follower or team member.

Even though the democratic leader makes the final call, every employee or follower has an equal say on the direction of the goal.

This democratic leadership style is usually effective because it allows all employees to have a say and exercise authority regardless of their level. For example, a democratic leader may give their team members different decision-related options in a board meeting and discuss each option before agreeing. Then they may agree by opening the decision up to a vote, or the leader may consider the team's feedback before making the final decision.

Pros

- Employees being listened to is an important aspect of the democratic leadership style
- Employees are involved in the business aspect of an organization.
- There is transparency as the leader shares details with the employees.

- The leader can easily gain respect, build trust and commitment, and drive responsibility and flexibility.
- The democratic leadership style can be very effective when the management is faced with a difficult decision to make.

Cons

- Reaching a consensus isn't always easy.
- It can require a series of meetings because ideas are mulled over without reaching an agreement.
- Not effective when employees aren't competent enough to offer good advice.
- Leaders sometimes use this style to stall making important decisions, leaving the employees feeling confused and without a leader.

2. Servant Leadership

The servant leadership style is when a leader puts the well-being and needs of their followers first. They use a serve-first approach in prioritizing their employees, organization, and companies before themselves.

Pros

- It encourages the growth and development of others.
- It increases the trust of leaders.
- It leads to innovation and improved performance and encourages collaboration.
- It reduces turnover and discourages disengagement.
- It encourages a safe space where people aren't scared to fail or make mistakes.

Cons

- It can be resourcefully intensive.
- The leader can experience constant burnout.
- The leader can be easily seen as weak.

- It can take longer to achieve goals or see certain results.
- It's difficult to encourage and train leaders in the serve-first mindset.

3. *Autocratic leadership style*

The autocratic leadership style is also referred to as the authoritarian leadership style. An autocratic leader primarily focuses on efficiency and results, regardless of what it takes. They make decisions independently or with a small group and expect every other member of the team or employees to do as they are told without questions. Autocratic leaders are also seen as military commanders. This leadership style is useful in organizations with strict guidelines or employees who need serious supervision (employees without much experience).

Pros

- It allows for efficient decision-making.
- Individual roles are clear; everyone is given specific duties and aren't

given an option to step out of their role.

- The team is more cohesive since only the leader is in charge.

Cons

- It prevents diversity in thoughts.
- It restricts creativity, innovation, and collaboration.
- It hinders professional growth and leaves no room for mentorship.
- It leads to low-spirited teams who feel they don't have a voice.

4. Coaching Leadership Style

The coaching leadership style is when leaders recognize their team members' motivations, strengths, and weaknesses and help them improve individually. The coaching leader helps team members set smart goals and gives regular feedback to promote their growth. They usually set clear expectations and create a motivating environment for employees to work in. This leadership style is beneficial to employers and employees. However, it is also an underused style because it is more time-consuming than other leadership styles.

Pros

- It encourages collaboration and two-way communication.
- It encourages being supportive and not judgmental.
- It allows constructive feedback.
- It encourages growth and creative thinking.
- It boosts the personal and professional growth of individuals.

Cons

- It doesn't always give the fastest and most efficient results.
- It can be resource-intensive as a result of the energy and time involved.
- It is sometimes not the ideal choice for results-driven organizations.

5. Transformational Leadership Style

This style is similar to the coaching leadership style, where the focus is on goal-setting, communication, and employee motivation. However, in this case, the transformational leader focuses on the organization's objectives.

The transformational leadership style is best for teams that can handle many tasks without constant supervision. The leader spends much of their time on the overarching goals.

Transformational leaders provide encouragement to their teams, have mutual respect, inspire them to achieve their goals, provide encouragement, and are usually creative.

Pros

- Team members have the autonomy to do their jobs.
- It encourages growth, creativity, and empathy among team members.

- It provides motivation for the whole team.
- It provides a safe space to encourage collaboration and build strong relationships.

Cons

- The leader is burdened with much pressure because they need to lead by example.
- It may not be the right fit for specific organizations (for example, bureaucracies).
- It may cause feelings of instability.

6. Bureaucratic Leadership Style

The bureaucratic leadership style is similar to the autocratic leadership style. With this style, they expect their team members to follow the rules exactly as written. There is little to no collaboration and creativity since everyone has a set of responsibilities.

Pros

- There is no favoritism.

- There are clear and specific expectations, roles, and responsibilities.
- It offers job security.
- There are highly visible regulations and processes.

Cons

- It hinders creativity, free-thinking, and innovation.
- It can create difficulty in responding to change.
- It is inefficient since everything has to go through a chain of command.
- Doesn't encourage collaboration or relationships among teams.
- Prevents personal or professional growth.

7. *Laissez-Faire or Hands-Off Leadership Style*

Laissez-faire leadership is also known as a hands-off leadership style. It is the opposite of the autocratic leadership style, where tasks are delegated to team members with little or no supervision while the leader has more time to dedicate to other projects. This leadership style is adopted by managers when the team members are well-trained, highly experienced, and do not need much supervision. However, productivity can be affected if the employees are confused about the leader's expectations or need constant motivation to perform well.

Pros

- There is a reduced fear of failing.
- It provides increased innovation and creativity.
- It instills a sense of independence among team members.
- It encourages individuals to practice individual leadership skills.
- It establishes trust between the leader and team members.

Cons

- It can lead to low productivity.
- It leads to confusion about their duties, roles, and responsibilities.
- Conflict may easily occur.
- It is not effective with an unmotivated and unskilled team.

Knowing your leadership style will help place you on the path toward becoming a great leader. Whether you manage a small or big team, your leadership style influences how your team works together to achieve a common goal.

So which of the leadership styles do you think successful leaders use? The answer is simply ALL!

The answer is ALL because research has revealed that leaders who have mastered four or more of these styles perform best. The idea is to switch between the styles and use the most appropriate one in any given situation. Combining these styles helps leaders achieve both their short-term and long-term goals and impacts and empowers their employees to achieve the organization's goals.

What Does Female Leadership Mean?

Is female leadership any different from male leadership? According to research, women have a distinctive leadership style that's different from their male counterparts. Women's leadership style is linked with the transformational leadership style, which aims to increase team members' morale, motivation, and job performance. Women tend to work with their team members to identify needed change, have a shared goal, and work toward it.

In recent times, women are gradually making their leadership presence felt in organizational politics, administration, and other aspects. Women are now ready to challenge and overcome the obstacles barring them from entering leadership positions even when they have the skills and talents to be there.

The participation of women in key leadership positions can be greatly seen in Latin America and the Caribbean. As of 2014, there were female presidents in Chile, Brazil, Argentina, and Costa Rica.

This trend has been growing fast as we now have a significant number of women in leadership positions.

It has been suggested that women are natural experts in managing crises and adapting to change. According to author Betina Rama, the ability to manage change and endure uncertainties is important to any organization's success, and these are features many women effortlessly possess.

The director of human resources of a consulting firm, Tiempo Real Amalia Vanoli, noted that women executives possess higher emotional intelligence. This is shown in how they ensure good working teams, motivate their teams, and never lose sight of the intended results.

Organizations that prefer women for certain positions have seen the benefits of having female leadership, and there is a strong policy supporting gender diversity put in place.

Research reveals that women experiment with changes in their behavior when they assume a leadership role. Some features that were not part of their character intensify and begin to appear stronger. They are more precise with decision-making and have a prompt discerning capacity. Generally, women take leadership as a true challenge and focus on the goal at hand.

Women leaders use horizontal leadership to ensure inclusiveness, encourage participation, and give information and power to their followers. The aim is to create and support group identities.

The Importance of Female Leadership

Regardless of its size, no organization can function properly without women's participation in leadership roles. Women bring a special kind of perspective into collaboration and healthy competition among team members.

Organizations led by women make effective decisions and deliver good results. The important qualities needed to be a good leader include connecting, collaborating, communicating, and empathizing with people. These are all feminine qualities that help build a sustainable future for any organization.

A survey by Boston Consulting Groups revealed that companies founded by women have better financial results (Abouzahr et al., 2018). It is evident that female leadership is crucial in increasing the pace of societal transformation in the workplace and beyond.

There should be gender parity in leadership because that's when true progress can happen. Female leadership is important because we women are passionate, resourceful, and very creative. We see plenty of opportunities in a world that often looks gloomy, and as a powerful agent of change, we are practical, authentic, innovative, strategic, and action-focused. We are adaptable, sociable, and thrive in relational environments.

We are team-oriented and collaborative. We bring diverse qualities, skills, and perspectives to leadership, leading differently. We don't act or think like men;

our differences are what make us thrive and achieve success.

We can't underestimate the importance of gender parity and diversity in leadership. As leaders, we seek to improve situations and drive results. We seek to encourage social, economic, and political progress for everyone. Breaking the ceiling is a phenomenon seen in different situations. For example, when an organization isn't doing well, women are quick to step in and turn things around, even when there is a high risk for failure.

Women worldwide need role models, and aspiring leaders need to understand this. They need to be inspired by the achievements of other women. As leaders, we all occupy an important role in mentoring aspiring female leaders, focusing on developing them, and supporting them to make a difference in the world.

As the first woman to serve as a Speaker in the United States House of Representatives, Nancy Pelosi once said, *"Be yourself... You are the only person who can make your unique contribution. Your authenticity is your strength–be you."*

To support this, Kamala Harris once said, *"You are powerful, and your voice matters. You're going to walk into many rooms in your life where you may be the only one who looks like you or who has had the experiences you've had. But you remember that when you are in those rooms, you are not alone. We are all in that room with you, applauding you. Cheering your voice. And just so proud of you. So you use that voice and be strong."*

Traits of Successful Women in the Workplace

While discussing the significant impact of female leadership in the workplace, let's digress a little and talk about something equally important: the traits of successful female leaders.

The following are traits of successful women in the workplace and beyond:

They believe in themselves and know they can succeed

To succeed in a competitive workplace or business, you must believe in yourself, your abilities, and your skills. According to the founder of USA Network, Kay Koplovitz, women need to be comfortable enough to think their way through and execute their ideas to get their desired outcome. She added that believing in oneself and confidence is key to success.

They don't play the pity card

How many successful women have you seen whining about their gender? Successful women don't see being female as a problem; they rather embrace their unique selves and appreciate the many things they can bring to the table. They will never blame their failures on the fact that they are women. Also, they don't see their gender as an automatic ticket to open closed doors. They know success comes from determination and hard work, and they do the needful.

The composed attitude of successful women inspires younger women to believe they can succeed. We can all overcome the glass ceiling with hard work, the right attitude, and methods.

While speaking in an interview, the first female managing director of TIME magazine, Nancy Gibbs, noted that *"I like the fact that glass ceilings are breaking all over. Probably very soon, it won't even be something anyone notices when you have a woman taking over one of these jobs."*

They care about how they look

The way you look says a lot about you. If you dress well, it exudes confidence, shows that you have things in control, and commands respect. Think about how Michelle Obama or Marissa Mayer dresses. You won't see them looking anything but stylish and pristine. Their image is well represented in how they look, and this is shown on the front pages of top-rated magazines.

As female leaders, you need to know the image you want to portray while handling the different aspects of your life. Overall, what you wear and how you look says a lot about you.

They take care of their body

When you have a high-powered career where there's a lot of pressure involved, it's easy to become stressed. The build-up of adrenaline can be relieved through exercise. It goes a long way in releasing stress and clearing your thoughts.

It is also important to eat healthy foods to improve your concentration and sharpen your mind. Other added bonuses of taking care of your body include keeping your blood pressure and cholesterol down and staying in shape.

They don't get overwhelmed with never-ending to-do lists

A successful woman will never get overwhelmed with the challenges thrown at her or the long lists of things she needs to take care of. If you want to be successful, you will learn how to focus on your tasks until they are completed. You make a to-do list and ensure you get to the end of it. You shouldn't see this as a problem as you strive to push yourself to see how much you can achieve.

They have a healthy balance between their work and life

A successful woman has a healthy work-life balance. This doesn't suggest that they leave work and head straight to the bar to party the night away. It means they don't allow their work to ruin other aspects of their life. This is achieved by prioritizing what's important and making time for their life outside work. This time is used to unwind, re-strategize, and refocus on what's important.

They remain calm under pressure

To make it in the tough world out there, you'll need to toughen up and be ready to stand the pressure. If you can't handle pressure, you won't achieve a lot. Since pressure is inevitable, find coping mechanisms that will help calm you down when the pressure is on. You need them to make you calm and think logically. You need to see solutions and not more problems; being composed is the way to go when in a difficult situation.

They support other women

Take a look at Madeleine Albright. She was the first woman to become US Secretary of State and has always advocated for women supporting women. According to her, *"There is a special place in hell for women who don't help other women."*

Successful women derive joy from seeing a fellow woman's achievement; they support their female counterparts and wish them well. They mentor other women who want to succeed and help them climb the corporate ladder.

They have passion for what they do

Do you know that if you don't truly care about what you do or have a genuine passion for it, you will never achieve true success?

Try to think of any successful woman who doesn't have a passion for what they do. Kamala Harris is known today because she has a passion for politics. Anna

Wintour, long-time editor-in-chief of Vogue magazine, is famous today because she is passionate about fashion. Burberry CEO Angela Ahrendts is obsessed with her brand and everything it stands for. I could go on to mention names of successful women who are passionate about what they do and not the financial benefits it brings. These women have dedicated their lives to their careers and are genuinely successful because they love what they do.

They have empathy and are compassionate

Women don't need to be rude or bossy to succeed. The media has portrayed female leaders as cold, harsh, and even emotionless, which is far from reality. A 2007 study published in the *Journal of Occupational and Organizational Psychology* revealed that female managers who are good at reading the nonverbal emotional cues of their subordinates and showing empathy were held in high regard, while others without the sensibilities were harshly judged.

They form strong bonds

Strong bonds and relationships are important to successful women regardless of whether it's a platonic friendship or a romantic relationship. Successful women usually have a small and close-knit group of friends that encourages them and provides advice and a listening ear when they need one.

They are risk-takers

Success isn't achieved by remaining in your comfort zone or staying average. You don't need to play it safe when it comes to achieving great things. Step out of your comfort zone and own it. According to the CEO of Meta Mark Zuckerburg, *"The biggest risk is not taking any risk… In a world that's changing really quickly, the only strategy that is guaranteed to fail is not taking risks."* Successful women always take calculated risks to push beyond boundaries and make their success known.

Former first lady of the United States Eleanor Roosevelt once said, *"What could we accomplish if we knew we could not fail?"* Successful women live by these words because fear will only limit them from achieving success. They won't know what's behind closed doors if they don't open them.

They lead by example

The most successful female leaders out there always lead by example. For decades, women have remained underdogs in the business sector. We have to constantly prove ourselves more than our male counterparts. We've never had it easy.

Leading by example has always been a good way to motivate teams. Living by the rule shows you aren't asking for too much from your subordinates. You aren't asking them to carry out a role you can't perform. This way, you will likely gain their respect while boosting your knowledge of a different aspect of your industry.

Finally, you need to step out today and dare to be so that you can win tomorrow! These traits of successful women weren't developed overnight, so don't be hard on yourself if you are aren't getting the same results right away. Take your time to study their traits and follow in their footsteps. Over time, you will win big and create the future you dream of.

Now that we have a good foundation with this first chapter, we can move to the next chapter, where we'll discuss the difference between management and leadership.

CHAPTER TWO: THE DIFFERENCE BETWEEN MANAGING AND LEADING PEOPLE

"Effective leadership is putting first things first. Effective management is discipline, carrying it out."

- Stephen Covey

Leadership and management are terms that have been used interchangeably. However, they are not exactly synonymous; therefore, it is important to understand the difference between them.

So what are the differences between leadership and management? First, we need to discuss both terms and understand their meanings before outlining the differences.

What Is Management?

Management is defined as the act of dealing with and handling situations, people, and things. This process of management may include planning, organizing, and coordinating to ensure a certain result is derived. Managing a team or situation in the workplace usually involves continuous reassessment and modification of results to measure productivity and boost results.

Management involves performing certain pre-planned tasks regularly with the help of your team members or subordinates. As a manager, you are responsible for carrying out management functions, which include planning, leading, organizing, and controlling. A manager can be a leader if they successfully carry out leadership roles and responsibilities, including providing guidance and inspiration, effectively communicating good or bad, and encouraging team members to increase productivity level.

However, not all managers can successfully fit into the shoes of a leader.

The responsibilities of a manager are outlined in the job description, and the subordinates follow them because of the job title. A manager aims to meet the organization's goals and doesn't focus on anything else. The title comes with authority and the responsibility of hiring, promoting, and rewarding employees based on their behavior and performance.

What Is Leadership?

We've extensively discussed leadership and different leadership styles in Chapter 1. However, we'll still be briefly defining the term here to refresh your memory and give you a better understanding of the differences between management and leadership.

Leadership is defined as behavior that supports a person or group of people in achieving a common goal. Leadership qualities include the ability to inspire, motivate, and encourage others to chase their dreams and see their vision through. Leadership is focused on driving results by creating and maintaining skilled teams rather than completing tasks through management.

Generally, a manager should possess the traits of a leader. As a leader, you need to create and implement strategies to build and sustain a competitive advantage against other organizations. An efficient organization that wants to achieve success needs strong leadership and management.

Most times, leadership and management overlap in their functions. While this may be true, the definitions above have shown different meanings and should never be used interchangeably. They have different characteristics and functions and require different skills sharing similarities. However, the differences are shown in some circumstances. For example, while some managers don't necessarily practice leadership, others lead without a managerial role.

Managers are appointed or elected within an organization based on their knowledge, technical skills, and expertise, while the main leadership skill is to inspire and influence people.

It's important to have both great managers and leaders in the workplace. Leaders help achieve an organization's vision and mission, while managers ensure that work is done and the teams align with the organization's goals.

Management vs. Leadership

In the following section, we'll be discussing the main differences between management and leadership.

Leaders Set Vision While Managers Follow It

Managers and leaders play different roles in creating and executing the vision and mission of an organization. Leaders can be seen as *visioners* because they have a clear vision of where they want their organization to be. Despite this, leaders are not the only ones responsible for making visions a reality. Even though leaders are responsible to communicate the company's vision, mission, and goals throughout the organization, managers are tasked with keeping the company aligned with the company's goals and core values.

Many employees have complained that their companies don't do a good job of communicating the company's goals, even though the manager attempts to motivate people to work toward a common goal. Meanwhile, employees expect that the details of the organization, how the organization is doing, and the direction it is headed should be communicated to them.

When the company's challenges, goals, and opportunities are made known, leaders work to build trust in the workplace. They ensure a productive working environment where the employees are willing to share their concerns, ideas, and needs. The transparency of the leader produces a healthier work environment.

Leaders Think of Ideas, While Managers Think of Execution

The manager focuses on control and rationality, while leaders care more about seeking opportunities to improve the level of their organization. This is carried out by generating new ideas and having a straightforward thinking mindset. In a nutshell, managers seek answers to "when and how," while leaders seek answers to "why and what."

The main duty of a manager is to carry out their tasks based on the vision of the leader they follow. They ensure that employees serve different functions and have different responsibilities to be productive, efficient, and heard. Managers control employees and give them the needed information, workflows, processes, and tools that encourage success.

A manager's relationship with others is according to their role in decision-making. On the other hand, leaders are more concerned with ideas and relate with people in more empathic ways, even when it's high-level. The difference here is the manager's attention to how things are done and the leader's attention to what needs to be done to give good results.

Leaders are always looking for new ideas while still actively driving change within their organization. Leaders inspire positive change by influencing and empowering their employees/team members to achieve a common goal. Effective communication is a leader's most powerful tool for ensuring this is done.

The way a leader communicates should get people ready to act differently and give reasons to support it, while the manager should be there to reinforce the messages that have been communicated. Many managers are not aware of their roles and the changes they are influencing.

Leaders Inspire People; Managers Drive Success

Leaders are tasked with inspiring people, while managers drive positive work experience and continued success throughout an employee's career.

Managers account for a larger percentage of employees' engagement and are also responsible for the productivity and success of their team. However, when employees are not inspired by the leader's words, managers are helpless, and there is little they can do to ensure success. With a personalized leadership style that encourages effective communication, self-reflection, and continuous feedback, leaders can empower employees, get their attention, and inspire them to go after their initiatives.

According to research, employees experience less stress and pressure when engaging with leaders regularly. They work efficiently in a workplace, supporting honest, transparent and open communication. However, many organizations ignore the importance of two-way communication between employees and leaders. The information is one-sided, leaving the employees out of company-wide conversations.

Leaders Look Ahead, While Managers Work in the Present

While leaders are more future-focused, managers focus solely on the present. As a result, the main goal of a manager is to achieve the organization's goals by implementing procedures and processes that will lead to success. These are staffing, budgeting, and organizational structuring. On the other hand, leaders engage in critical thinking and explore future opportunities. The leader's vision will amount to nothing if it is not clearly communicated to the manager and employees.

A leader should create a sense of purpose among employees through employee engagement and alignment of employees' personal and professional values.

Leaders Shape the Culture, While Managers Approves It

When comparing managers and leaders, every organization's corporate culture is an important aspect to look into. Culture is the system of beliefs, behaviors, and values shaping and defining how an organization operates to achieve its goals. Organizational culture should be aligned with the employees, business strategy, and the actions of other stakeholders supporting the achievement of business goals.

The difference between management and leadership in organizational culture is that while leaders define and shape the culture, managers lead their employees to create it. The leaders uphold the core beliefs and values of the organization through communication, action, and decisions. Their leadership skill and style influence how employees accept and live in the organizational culture, while the manager shows continual support and endorsement for the culture within the team. Therefore, a collaboration between management and leadership is essential to drive employees to uphold the organization's culture and core values.

Skills of Effective Leadership

It requires more than the ability to succeed to be a good leader. A survey conducted by the Harvard Kennedy School Center for Public Leadership revealed specific qualities that inspire people to follow a leader and have confidence in their decisions.

Cultivating strong leadership skills is what promotes certain behaviors and practices needed for achieving your vision.

Some of these skills include:

Integrity

A great leader shows great integrity and inspires people to trust her. Trust is needed for people to follow you, whether as their leader or manager. You can act with integrity by:

- Keeping to your word.
- Acting with consistency, regardless of your audience.
- Taking responsibility for your words and actions, even if it means admitting to making mistakes.
- Being transparent and never hiding your actions from others.

Visionary

Effective leaders know where they want to be, where they stand, and they involve others in charting the organization's direction.

Inspire Others

Leaders are naturally inspirational by helping others understand their roles in the bigger world.

Good Communication Skills

Leaders are good communicators that keep their teams updated on what is happening and what will happen in the future. They also communicate possible obstacles they may face.

Ability to Challenge

Leaders can challenge the status quo; they don't settle for regular. They usually think outside the box regarding problem-solving and creative thinking.

Important Skills for Successful Leadership

There is no leadership without followers. This means that a leader needs followers to be able to lead. Therefore, skills for managing and leading people are important for successful leadership.

The following section will discuss the three important skills leaders need for successful leadership.

Motivating Others

The key to leadership is to motivate others to strive for what you want. You can motivate your team by providing them with an area of interest, setting challenging goals, and stimulating work. A motivational environment can also help improve motivation.

The three elements that usually work for motivation are: showing people how much you appreciate them, encouraging them to develop their skills, and helping them see the bigger picture.

Delegating Work

Delegating tasks and achieving success from the task delegated is important for successful leadership. As a leader, you can't do everything yourself. If you attempt it, you may find yourself struggling, and it can even cost you your team.

The key to successful delegation in leadership includes understanding the type of control you want and communicating it clearly to your team members. Success can only come when both parties understand each other.

Facilitation

Facilitation skills are important for project management, change management, and team building. The idea is to help both the leader and team members identify their goals and achieve them.

Finally, good leadership leads to increased morale, motivation, and productivity in every organization. Every leader needs followers who share their ideas and join them to achieve a common goal. Motivating and managing people are key leadership skills you need to develop on this journey. Your leadership will fail if you can't bring others along with you.

CHAPTER THREE: MANAGING DIFFICULT PEOPLE AND SITUATIONS TO SUCCEED

"Gossip and drama end at a wise person's ears. Be wise. Seek to understand before you attempt to judge. Use your judgment not as a weapon for putting others down, but as a tool for making positive choices that help you build your own character."

- Marc Chernoff

As a leader, managing people is, by a long shot, the hardest obligation of all. Each individual is different, free, and moved by entirely unique factors. While having many people at hand can make your team achieve more, every leader will ultimately experience issues from difficult people. However, when you face this issue, don't be dismayed—it's not a new problem.

Even as unpleasant as it is dealing with difficult people, this management challenge is worth the trouble. Why? Because you are facing what other leaders and managers are facing—or have faced. This means that there is a way out.

An important question you may want to ask is: *What does a difficult employee look like?* It's imperative to recognize the traits in your subordinates that will tip you off that they are difficult.

What Does a Difficult Employee Look Like?

A difficult employee does nothing but contribute negatively to your team. They cause you and your team mental stress and low morale. At the end of the day, your team may end up yielding less output than expected.

Remember that identifying a difficult employee is as important as your leadership role itself. Identifying their traits will allow you to address the problems swiftly and help prevent future occurrences.

Below are three behavioral traits of a difficult employee.

Irresponsibility: The employee doesn't perform as expected

Sometimes, you will face employees who seem like they are not serious about fulfilling responsibilities. This poor employee performance can result from the absence of inspiration, capacity, or both. It could also be due to poor communication, the absence of assets, or a unique set of factors.

You should be concerned when a subordinate doesn't show the zeal to work. Don't be quick to conclude that their lackluster attitude is due to laziness. There is always a reason for every action.

Mental health experts have revealed that employees who expect more and receive less tend to show lackadaisical attitudes toward their work as this works them up mentally. Imagine a scenario where Mr. A is capable and qualified to handle the company's finances as a financial secretary, but he's being put in charge of staff records. There is a 70 percent probability that he won't like his job. As a result, he might not give his best to the role.

Personal life also matters. Not every poor employee's performance results from the job. Some employees are facing numerous problems under their suits. Not everyone has the same tolerance for personal issues, and perseverance levels differ. They might be facing challenges at home or in their relationships or health that can affect their performance at work and reduce their efficiency.

Despite the genuine reasons they could have, poor performance at work isn't something you should condone for a long time.

Bad Behavior: The employee exhibits unacceptable behavior

Some employees are naturally troublesome. They will give you nothing more than headaches, energy drains, and create low morale in their team.

Employees with bad attitudes often don't show it directly; they rather show it when you aren't there. However, this trait can be seen in employees through different means, including:

- Eye rolls or smirking
- Being distracted during meetings
- Getting to work late
- Being skeptical about everything
- Looking down on the efforts of co-workers

While a bad employee attitude can be entertaining in the initial stages, in the end, it is always damaging. It's disruptive and should be curbed as soon as it is noticed.

You may be wondering why it should be addressed quickly. This is because those kinds of employees have negative energy. If they are on your team, they will make the working atmosphere unbearable in the long run. As a result, misunderstandings may set in among the teammates or employees, leading to poor work habits and lower productivity.

Also, if a difficult employee has a similar bad disposition while managing customers, your business and reputation are in grave danger.

Power Problem: The employee undermines your authority

Control is essential when performing your role as a leader. You need to be able to give orders and receive obedience from others. An employee who defies your leadership by undermining your authority upsets the work environment. This behavior can convince other employees that you are not capable as a leader.

Some employees undermine your authority unintentionally and not with destructive goals in mind. They might have a unique reasons and viewpoint that you are not aware of and fail to see how they relate and undermine others.

On the other hand, an employee's bad conduct might be a conscious endeavor to sabotage you. That's why as a leader, you need to be able to determine their motivation. This will help you choose the best way to address the problem.

Dealing with Difficult People

A huge part of dealing with difficult people depends on your level of knowledge and experience. But this is just a fraction of what needs to be done. Being a good leader and manager requires you to responsibly perform your "roles."

You should ask yourself: *"What do I need to do?"* Or better still, *"What am I doing wrong that I need to change?"* If you can provide tangible answers to those questions, you're on your way to becoming a good leader and an excellent manager. Only then can you know how to deal with troublesome people—particularly your employees.

Tips To Effectively Handle Difficult People

Here are some tips on how to effectively handle difficult people.

Criticize the attitude, not the person

Here the point is to address the unacceptable behavior without making it personal. The aim of trying to handle a difficult employee is to know the true cause of their "being difficult."

In order to be the good leader that you can be, you should assist your employee in becoming better. You do this by showing concern toward their attitude, not making them feel bad about themselves. You're a problem solver, not a critic—remember that!

Call their attention to their behaviors and listen to them attentively. Hearing them out will give you the information you need about them and allow you to find the best way to address the problem. Get them counseling, if possible. Be a motivator—inspire them!

You can also let them know what you've been noticing about them—in an interactive manner. However, try to do this in a way that they won't be offended.

Find out the reasons behind their "being difficult"

Whether it's obvious or not, every employee looks up to you like the general overseer of the whole team. This implies that it's part of your responsibility to determine the real causes of your employees' behaviors.

Sometimes, employees behave in difficult ways without understanding the impact of their actions on the whole team. The reasons behind their actions can be anything from family problems to workplace issues, or even more!

Common reasons for employees' bad behaviors include:

- Death of loved ones
- Marital issues like divorce and single parenting
- Heartbreaks or relationship failure
- Co-worker issues
- Unbalanced reward for services rendered
- Health issues

Therefore, as a leader, do your best to get to the root of the problem. This will give you the reputation of a leader who understands what needs to be done.

Be approachable

A lot of leaders are the actual cause of difficult employee behaviors. Most of them are locked up in their offices and only respond to messages through their secretaries. This makes employees feel that they're not open to getting feedback from people.

Be open! Be approachable! Be open to deep conversations with employees. Not only will this enlighten you about them, but it also makes them see you as a leader they can confide in.

More importantly, give employees room to provide feedback on decisions made, the output produced, customer relations, and assigned roles. Being approachable builds trust between you and your employees.

Let your guidelines be understandable

Setting clear, straightforward instructions can help deal with employees' difficult behaviors. People do better when the instructions given to them are simple, clear, and understandable.

If you're trying to instruct Mr. A (who's working in the financial department) that he should report to you with the team's financial reports, get straight to the point by letting him know what exactly he has to do.

Hence, be more specific in instructing others while making the instructions understandable. This can save the team from poor employee performance.

Write down goals and consequences if not met

You need to write down your expectations from your employees. Using sticky notes doesn't make you look inexperienced or disorganized. Instead, it shows that you're a goal-getter.

Come up with a plan containing details about the goal, how it flows toward the project's success, and proposed deadlines. You can also write out how not achieving those goals will affect you, other employees, and your role as a leader. Then hand the goals over to him. The right mindset will make him more serious and focused, and that will benefit the whole team.

Supervise and track employee's progress

Usually, this comes after goals have been set. It's a way to track assigned projects, perceive the level of employees' dedication to assigned projects, and check progress rates. Also, this helps ensure that the employees complete assigned jobs within the timeframe given.

You might ask: *"How can I check a job's progress?"*

You can do this by:

- Having a roundtable meeting where everybody can contribute to the progress of the jobs at the employee's hand.
- Encouraging employees to send in their thoughts about the project and its success.
- Sending personal invites to employees for one-on-one chitchats. Here you can learn more about the methods used by each employee in getting the job done.

Write down problems posed by employees: Documentation

While it might seem unnecessary, you can also write down the behaviors you notice about your employee. Tell them during a private meeting. There you can talk about each trait you've noticed and the way forward. Bring it to their notice that they need to change.

Keeping a record of problems posed by an employee means having evidence. If Mr. A still refuses to change after your dialogue, you reserve the right to let him go. In a bid to sustain his job, Mr. A might decide to sue. You can win the case if you have evidence of his past conduct.

Adopt Preventive Measures: Planning

Prevention, they say, is better than cure. If you're going to hire a new employee, run a thorough background check on them. Ask them questions about their former workplace, why they left, behavior with customers, etc., and pay close attention to their reactions.

Just as other forms of interviews are important, don't forget to conduct a behavioral interview on your candidate. Do all you can to ensure that you get the best candidate for your team. Being careful allows you to be aware of red flags and address them immediately.

Respect is Reciprocal

Don't misuse your position as a leader to disrespect your employee. Even when it comes to the firing stage, you don't need to burst out with all elements of anger in you. No!

Be composed, professional, and relaxed. Avoid making unnecessary comments, criticisms, and assumptions. This is to avoid creating a scene or a bad reputation as a leader.

Fix your attention on what you've got. Let your tone be serious but not harsh. Also, make sure you listen to your employee's side of the story. Firing someone isn't always the best option.

When you talk calmly to your employee, you automatically command a calm reply from them. A difficult employee is just difficult and not a criminal. So try to address them respectfully, and you will get respect in return.

Solutions for Dealing with Difficult People: The Way Out

There's always a way out of things when dealing with challenging employees. You should note that working with difficult people can deplete productivity and establish an unfriendly workspace for others. That's why you should step up your management game by taking swift, necessary actions and exercising authority when due.

Examining employees' conduct, performances at work, making precise arrangements to address employee problems, personal engagements with employees, and supervising troublesome employees' progress are great ways to deal with the problem.

Sometimes, third parties can be involved when the problems become serious. These third parties are usually units with the sole aim of attending to people management issues. You can engage them to make your employee better rather than being difficult.

Here are some tips for solving the issues of difficult people.

Separate the Troublesome Employee from the Team

According to Harvard Business Review, bad employee conduct can spread across the whole workforce if care is not taken. If you cannot successfully address a difficult person or cannot immediately terminate their appointment, separating them from the team is a good option. Doing this can help teach others a lesson, curb misconduct, and save the team from a toxic atmosphere.

It might be hard to directly separate an employee from the rest of the team. You don't want to send negative mental signals, and at the same time, you can't afford to risk your team's productivity.

You can consider doing any of the following to separate difficult employees from the rest of the team:

- Give more work to other employees to get more work off the difficult one.
- Reduce the frequency of team meetings.

- Re-shuffle workstations to isolate a difficult employee.

Engage Human Resources

Consulting the Human Resources department is another way of finding a solution. They specialize in personnel management, and as such, the department is a great fit for your employee-related matters. They can assist you with recording employee conduct, the things you've done to deal with that conducts, and termination procedures. They are versed in organizational policies in dealing with difficult employees.

Be a Problem Solver

Ideally, a leader is expected to function as a problem solver—and you can become one. Consider your role in the problem's solution. The goal is to bring a better person out of your employee. That's it!

Try personal methods in bringing the best out of your employees. That's what makes you a great leader. Also, work with the employees in question and other employees in addressing the problem.

Put Yourself in Their Shoes

Sometimes, the best way to solve someone else's problem is to assume their position. Assume you're the problematic employee and ask yourself:

- What could make me give my boss a tough time at work?
- Am I given more or less work than necessary?
- Am I being appreciated enough?

Once you put yourself into their shoes, you may be able to identify what causes their lack of motivation and address it as soon as possible.

Some employees want to do more, but they're given less work, and they feel that you don't recognize their impact. Others are overwhelmed by the workload, and you don't notice. In other cases, what employees get as rewards or appreciation is nothing compared to the tons of work they do.

If you can highlight these situations, you're good to go—on your rescue mission.

Employee Development Programs (EDPs)

Getting your employees to enroll for EDPs is a great idea in dealing with difficult employees. Make the arrangements for ALL your employees to enroll. Avoid enrolling just the difficult ones to avoid creating stigmatization.

Be updated in seminars, meetings, or webinars on people-related issues and relay the information to your employees. The team's success depends on them, and that's why you need to invest in them.

EDPs will help to "unshape," shape, and reshape your employees' mindsets toward work and workplace authorities.

The Last Resort: Letting Them Go

Some employees are real thorns in the flesh. They can make work unpleasant for others and leadership ineffective for you. Even though you should usually be cool, understanding, and sympathetic, there are times when you have to be firm.

Being firm doesn't mean being harsh. It means maintaining your stand that you've had enough. Don't forget that you're the boss, and you decide who stays.

Of course, nobody wants to create problems for others, but when you've had enough, it's enough. Don't allow emotions to keep you from doing what needs to be done. Point out why you're terminating their employment and state what you've done to address the problem.

Also, let them know that you've been studying them even after calling them to order, but they haven't made amends. Ensure that you talk to the Human Resources department for due process in the employment termination.

If you fail to take drastic actions when necessary, you're putting your organization and workforce in danger of becoming a toxic environment.

CHAPTER FOUR: OVERCOMING SELF-DOUBT AND PROVEN TECHNIQUES

"The worst enemy to creativity is self-doubt."

- Sylvia Plath

Normally, we all feel self-doubt when confronted with new or strange circumstances. At that point, we feel that we're vulnerable to anything. We feel incapable of doing what we actually can. This is something we will all face at some point in our lives. Notwithstanding, when it becomes unbearable for us, that's the point at which we might require more to eliminate that feeling.

This feeling might originate from past bad encounters or dealings with others. There are a lot of factors that can contribute to that feeling. For instance, Miss B has been previously told that she's not "a great fit" for a job. At that point, there's a high chance of her doubting herself, and her self-esteem can suffer. The fact that we are being pressurized on all sides to be "societally acceptable" also makes this feeling inescapable.

What exactly is self-doubt? Without further ado, let's get into it.

What Is Self-doubt?

I'm sure you've met people who are very bold and confident. You might even admire their solid body language, their eloquence, and the way they carry people along when addressing them. Whenever they enter a place, everyone notices. Whenever they clear their throats, everyone pays rapt attention. You might think these individuals have always been like that or were born like that. But the truth is, even the most confident people struggle with self-doubt.

Self-doubt is the lack of confidence or trust in yourself and what you can do. It is the feeling that makes you believe that you can't do what you want/need to do.

We're humans. We won't always be sure of ourselves. However, how we manage these facts determines the difference between the goal-getters and the regular visionaries.

Like Tony Robbins said: *"I don't have to get rid of the fear; I just have to dance with it."* Successful individuals have dwelled on this fact and converted their self-doubts into motivations to soar higher and be more grounded, and you can do that too.

Self-doubt is nature's way of protecting you from loss, disappointments, or being laughed at. However, instead of avoiding embarrassment, many people simply lose opportunities. If you keep allowing self-doubt to control you and hinder you from taking bold steps to maximize your potential, you're basically undermining yourself.

Another question that might interest you is, *"What causes self-doubt?"*

What Causes Self-doubt?

Self-doubt arises when we need assurance or feel unequipped to do the things we really want to do. People who are uncertain about themselves experience vulnerability around things they can't handle. They also tend to stress over things when they don't work out as planned.

A small degree of self-doubt is great since it shows that you understand what you need to work on. It also shows that your intentions are clear, positive, and big. Nonetheless, fear and doubts can influence your life in a way you don't want.

The following are common causes of self-doubt:

Previous Experiences and Mistakes

Our past encounters can massively affect how we respond, particularly bad ones. Bad experiences like unhealthy relationships or being unjustly fired at work can drastically affect us mentally.

You can get disoriented as a result of your previous encounters. However, you can become a better person by gaining from those experiences instead of letting them affect you. This is to ensure that you make the best out of future situations.

A Competitive Mindset

Since we live in a competitive world, it's normal to compare ourselves with other people. We all want to be the best in what we do. In fact, the media has influenced us to the extent that we fear that others are better than us. Nowadays, it's as simple as going online, seeing other people's looks or lifestyles, and drawing a conclusion that they're better than us. You might begin to feel like you're working less than you should, or you're not doing the right thing. When this competitive spirit gets into you, you begin to feel uncertain about your real essence.

Poor Formative Years

The way we spend our formative years heavily shapes our character and views. The people you spent your early years with are factors that can determine whether you'll be a confident person or not.

Studies have shown that people brought up under guardians who trained them to satisfy others grow with self-doubt in them. If the school you attended judges students mainly on grades, self-doubt is also bound to set in.

New Encounters

It's natural for you to feel uncertain when faced with a new or entirely different challenge. It's because you've never been faced with a situation like that before and are unsure how to respond in that case.

You might have just been promoted to a higher post or qualified to a new stage in a competition. The feeling of doubt and not being equipped might not be comfortable. That's self-doubt!

It makes you ask yourself questions. Instead of trusting your abilities, you begin to look blindly at situations. Hence, you need to brace yourself in preparing or handling new challenges.

How Self-doubt Affects Your Performance

Does self-doubt affect your performance as a leader? Of course!

You should note that your performance determines your output, and your output speaks either good or bad about you as a leader. Therefore, it's imperative to know the impact of self-doubt on your performance because once self-doubt sets in, there's absolutely no aspect of your life that it can't affect.

Business

It's very easy for self-doubt to affect your business. Imagine a scenario where you are an entrepreneur, but you lack good communication skills. You may have excellent business ideas, but self-doubt sets in when talking to key persons who can influence your business.

You'll feel disbelief in yourself and eventually not make a move. Hence, you've not only failed to break your personal boundary, but you've also failed to perform well in your business.

However, a little self-doubt is necessary. It keeps you from being overconfident, which can also affect your business negatively.

Leadership Roles

At times, we all battle with self-doubt. However, that battle can be particularly challenging for leaders. Leaders are generally bold, outspoken, and confident in today's society. But, as a leader, what happens when you start to not trust your abilities?

From the fear of forgetting what to say at team meetings to the *"Can I?" "Am I?" or "Will I?"* that can set in when you assume a new leadership role, self-doubt has a huge impact on your ability to lead.

Okay, you forgot some important information at a team meeting. You felt so ashamed of yourself that all you wanted to do was go back to your workstation, pack your things, and never come back. Unfortunately, you're the leader. You might start to think you're not fit, or you shouldn't have accepted that position.

If you keep the doubts rolling in, you're doing yourself no good as a leader. You will end up making bad decisions which, of course, means you've performed poorly.

Self-doubt: A Killer-Feeling-Turned-Advantage!

As a leader, you can master the art of using self-doubt to your own advantage. Why? Because it has actually done more harm than good. You must pay careful attention when studying parts of yourself that you're doubtful about. There'll be times when you will be uncertain about your role as a leader. You might ask yourself questions like: *"Am I really fit for this?"* or *"What if I don't do well?"*

Whenever these scenarios start coming up in your head, don't think you're crazy. No! It's actually an interaction between your mind and your brain on the safest way to run things. It's a trying-to-play-safe scenario.

A leader that doesn't have doubts is in big trouble. Lack of self-doubt is the beginning of overconfidence, leading to failure in the long run. Although fear of failure, employees, etc., can instill negativity into you, you can still channel that self-doubt into a useful weapon as a leader.

Here are some tips:

Identify What the Doubt Is About

Whenever you feel doubtful or scared, try to calm yourself down. This will give you enough space to begin your search for ideas. Self-doubt comes alive whenever a particular thing has to be done. This implies that self-doubt is a feeling based on specific necessary actions.

Hence, try to identify what triggered the doubt. It could be your eloquence at a team gathering or a decision you're about to take. Whatever triggered the fear, find it out.

Work Toward Quenching the Doubt

After you've succeeded in identifying what the doubt is about, work toward that thing.

If it's based on people's opinion of you, you may need to think harder because whatever decisions you make are bound to impact others. The main goal of this is to make you a better decision-maker and not a fearful leader.

Quenching that doubt makes you free of fear. A leader has fears, but how you address yours makes you unique, outstanding, and successful.

Cross-check Your Feelings

After all is said and done, check within yourself again. Are you still feeling as doubtful as before? Is that feeling suppressed? Is the feeling more than before?

If you're feeling as doubtful as before, it may be that you've addressed the wrong problem or you've made a wrong or unnecessary decision. If not, you're on the right path. Here, the good news is that your doubts are either suppressed or gone totally.

Self-doubt is not something that goes easily. You need to master your feelings and know the best method to manipulate your fears so that they will work as your friend.

Management

Just as self-doubt affects your role as a leader, it does the same in your management roles. Many managers do not achieve much due to self-doubt stopping them from moving forward. Others, including achievers, give credit to luck, chance, or other factors because they disbelieve in themselves and their abilities. This is known as imposter syndrome.

Successful females are usually considered to be the major victims of this syndrome. However, research has since shown that it occurs in both genders.

Self-doubt belittles a manager in a very real sense. You might fear that you will falter when giving out instructions or during a presentation. This deprives you of good communication skills. You'll feel drawn back when you should be at the frontline of various adventures in management. In the long run, you'll fail to deliver your duty as expected. Hence, you should take bold steps to overcome self-doubt to be a better manager.

How to Overcome Self-doubt

Allowing self-doubt to take over you, the decisions you make, or your dealings with other people is as risky as putting an incapable person in the care of your business. If care is not taken, you'll be sabotaging your career all by yourself.

We all know that self-doubt arises when you give in to the sensation of fear and uncertainty that you can't get the job done. But self-doubt can also arise from past experiences. Perhaps you were disqualified from something you really wanted, or you were voted out of a competition. Don't give in to that feeling. You're bigger than that feeling—self-doubt!

Overcoming self-doubt is like a game of soccer. When you're on the pitch, you'll be uncertain about hitting the back of your opponents' net. That's how it works.

But as you strive to get the ball, toss it around to your teammates, and receive the ball back, you start to feel positive about your goal on the pitch.

Although you might have fired some entertaining shots to no avail, your resilience will eventually pay off when you score a goal. And all that occurred because of one thing: your self-will!

There are a lot of ways by which you can overcome self-doubt, and they include:

Know that You're Not Alone

The first thing is to know that you're not the only one in that situation. There are millions of people in the same situation, and even worse. When you realize that you're not alone, improving yourself will not be a difficult task.

Believe me when I tell you that all leaders —even the most outstanding ones—have times when they sink deep into their shells because of self-doubt.

Knowing that you're not alone will allow you to embrace self-doubt as a part of human nature. Rather than getting yourself worked up, make it your friend. Look at other leaders, listen to their stories, and learn from them. Most successful people claim that their breakdowns led to their breakthroughs; you can also experience that.

Activate Your Self-will

Confident leaders aren't born; instead, they are made. Look at your role models and see how they run their businesses. They weren't born with confidence; they built it over time. Self-will brings about determination toward achieving your objectives. So to overcome self-doubt, include self-will in your plan.

Self-will allows you to look above obstacles and see the silver lining behind every cloud. It encourages focus, discipline, and passion for surmounting your problems. It helps you see your uncertainties as crucial parts of your success story. You may not like the situation you're in right now, but once your overcoming spirit is strong, the future is bright.

You can activate your self-will through the following:

- Believe in yourself.
- Ignore others' opinions about you.
- Don't go through it alone. Surround yourself with people with positive mindsets who believe in you.
- Encourage yourself daily with devotionals and inspirational materials.
- Change yourself—your viewpoints, attitudes, and the people you associate with.
- Get a guardian angel. This can be a family member, friend, colleague, role model, or therapist. They will mentor you on the necessary steps to take and how to take them.
- Watch your role models.

Relive Your Past

While it's healthy to let the past go, sometimes you need to find your past memories, relive them, and make yourself happy.

Do you remember that time when you won that award? Or the time when you got that promotion? Perhaps you once did a jaw-dropping presentation; remember that time. Don't forget that round of applause that you got from your employees. Remember the hearty congratulations and the feeling of fulfillment that flowed in you. Remember your favorite eating spot that you took yourself to. Remember that party you threw to celebrate.

Don't just smile alone. Remind yourself that you achieved great things with your positive mindset, hard work, and focus. Now tell yourself that you believe in yourself. Encourage yourself by saying: *"If I've done it before, I can do it again."*

Befriend Your Inner Critic

Every feeling of doubt comes from a voice within. This voice is your inner critic. It's what tells you that you're not worth that position, winning those awards, or getting that task completed. Your inner critic tricks your mind into believing that you have to play it safe every time.

But great leaders are known for the ability to take risks. So it's up to you to deal with that voice. You should note that you can't totally silence your inner critic, but you can make it your friend. This way, you can seize the control it has over you.

As a friend, you can give it a name. Let's say you name it Lan. Anytime you feel uncertain about anything, you can say, "Hey, Lan, I've got this one." Better still, you can consult it for potential lapses in what you want to do. You can say, "Lan, in today's presentation, what are you scared of?"

Just make sure you're in control and not your inner critic.

Set Short-term Goals

You might decide to talk to someone about your situation. You might decide to start doing yoga. You might want to write, take up a sport, or implement a new habit at the office. Whatever it is, try to put it in your daily schedule.

Accomplishing these short-term goals will give you a sense of happiness and fulfillment. This will help in boosting your self-confidence and quench your self-doubt. After a while, you'll be radiating positivity everywhere you go.

Be Selective with Feedback

Not everyone has the same positive mindset as you. Knowing this will help you in filtering out the feedback that will boost your self-confidence and the kind that will make you doubt yourself. However, you don't necessarily want to disregard negative feedback. Sometimes it is just constructive criticism and can be useful.

Forgive Yourself

No method of overcoming self-doubt can work if you don't work on and with yourself. You have to forgive yourself for doubting yourself. Forgive yourself for making mistakes. Forgive yourself for not looking at the bright side of your situation.

Think of how you can be better. Make peace with yourself. Think of ways by which your obstacles can be surmounted. You can also go back in time to see what you should have done when you doubted yourself. This way, you can recover your losses.

CHAPTER FIVE: HOW TO BUILD HIGH-PERFORMING TEAMS USING COACHING AND DEVELOPMENT

"Teamwork is the ability to work together toward a common vision. The ability to direct individual accomplishments toward organizational objectives. It is the fuel that allows common people to attain uncommon results."

- Andrew Carnegie

In today's world, the foundation of successful work environments depends on high-performing teams. Every team has its own function. For instance, the leadership team runs partnership deals with other organizations. Production teams introduce new items to the market.

Network teams are in charge of helping other units. They assist in reaching out to other organizations, sourcing and conveying information for the organization.

Amazing teams are a must-have for all organizations. How they organize their proposed actions, distribute work, and combine ideas is appreciable. This unique way of operating ensures that the outcomes of work done by such teams are second to none.

A team is the coming together of two or more people to cooperate and achieve a common goal. The common goal is more important than personal interests, and members need to feel responsible together.

You might have a question: *"What is a high-performing team?"* Here, you'll understand the concept behind high-performing teams and their qualities. You need to know the most effective ways of building high-performing teams and managing them in your organization. You also need to know what you need to achieve this.

You'll learn how to build and manage a high-performing team.

What Is a High-performing Team?

You'll agree with me that teamwork experiences can be unique. At some point in your life, you've found yourself working in teams that you didn't gel with. You've been in teams where the working atmosphere was not pleasant. These experiences create continuous anxiety when you don't know what misfortunes await you at work.

Like Reid Hoffman said: *"Regardless of how splendid your brain or system, in the event that you're playing a solo game, you'll constantly miss out on a group."*

A high-performing team consists of qualified, talented individuals with relevant abilities who are highly attentive, ready, and focused on accomplishing clear, extraordinary outcomes. They do things in styles that outmatch normal performing teams.

High-performing teams work toward outstanding results through excellent communication, team trust, unique aims, effective task distribution, and positive arguments. Each team member claims responsibility for their own tasks and actions.

What separates high-performing teams from the normal performing teams is that the former is a collection of individuals who make conscious efforts toward achieving the team's goal.

So what makes a high-performing team what it is? It is simply the members of the team. Let's look at some traits of members of a high-performing team.

Members of a High-Performing Team

A high-performing team is what it is because of its members, as they are the decision-makers and executors. Thus, it'll still fall on their desk if anything goes wrong. Look at top businesses; they have highly successful individuals that make up a team of leaders, a team of production experts, or a team of marketers. In a high-performing team, members have the following traits:

- Profound feelings of direction and obligation to the teammates and the team's goal
- A good understanding that they need each other to grow. Hence, members' shortcomings are not criticized. Instead, they're improved.
- A vast scope of skills that outmatches normal teams
- Team spirit, trust, and effective communication
- Dedication towards achieving high performances than members of normal teams.

It's almost inescapable to use teams at workplaces all over the world. Top-notch leaders can testify to their membership in very productive teams. We can say that a leader needs a team for personal development. High-performing teams enjoy smooth sails more than normal performing teams because everybody's contributions matter. Members can freely bring new suggestions, skills, and perspectives to the table.

Furthermore, with due diligence, high-performing teams have clear guidelines that members need to follow. Members often beat deadlines for assigned tasks and might not need strict supervision since they're more than committed and responsible for execution.

Since humans have to be constantly motivated to do better, members are regularly motivated through gifts, remuneration, and awards. Also, the team's aim is to be a top-notch team; poor performances are not encouraged.

Characteristics of High-performing Teams

There are a lot of characteristics peculiar to high-performing teams, most of which are worthy of emulation. According to Professor Ina Toegel, a high-performing team should not be more than eight people. This number or less means all members are easier to coordinate. It also means there'll be no case of "too many cooks spoil the broth."

Below are some characteristics of high-performing teams.

They pursue clear, straightforward objectives.

High-performing teams have members who are serious about the team's achievements. You can see that in their concentration at work and their search for more ideas to improve the team. Not only are their goals clear, but they're also achievable. In some cases, those goals might seem impossible.

To give life to their collective goal, members search for more ideas. The ability to think outside the box assists their work in getting accomplishment. The team's objectives are clear enough, and duties are shared among teammates so that everyone has a part to play. This way, each member knows what to bring to the table.

They Know Their Place in the Organization

In the long run, the organization's mission is the ultimate goal. So all teams try to work toward that goal. A good understanding of how the team's goals align with the organization's objectives makes high-performing teams active and productive. They know why and what they want. Hence, they collaborate in achieving that common goal.

Their Meetings Are Strategic

Undoubtedly, inappropriately planned team meetings make members feel frustrated, tired, and uninterested. As a result, lesser productivity is bound to be achieved.

Experts have revealed that high-performing teams hold strategic meetings necessary for planning, innovations, and all-around development. They also put up ways to cultivate more useful social affairs. They host meetings on the premise that time spent together can be productive and worthy. However, these meetings also serve as doors to other profitable connections, creating healthy relationships.

They are not ignorant of their duties

Negligence of duties can lower a team's productivity and cause misunderstanding among members.

Misunderstanding can rapidly wreck a generally capable and useful team. High-performing teams prevent misunderstandings by clearly highlighting each member's role. This keeps the work environment clear of confusion about who is to do what. In short, responsibility is assured as high-performing teams always keep members, jobs, and deadlines coordinated.

They Value Effective Communication

High-performing teams know the essence of good and effective communication. This shows how they interact, exchange ideas, and show respect. They know that misunderstandings will arise when effective communication is lacking, leading to low team performance.

To ensure good communication, high-performing teams put in place necessary tools. These tools will facilitate the exchange of information and ensure that members communicate with the right teammate at the right time.

Conflict is a crucial part of a team. It helps to see a problem from different viewpoints. However, high-performing teams are skilled in manipulating conflict to their advantage—and perhaps stop similar problems from arising in the future.

They Use Scale of Preference and Practice Effective Time Management

In economics and business management, the scale of preference is a to-do list. The most important plans are placed at the top of the list, and the list goes down in decreasing order of priority.

High-performing teams make use of this in ensuring that first things come first. They first attend to the most important tasks and spend their well-planned time on them. They are aware that all tasks don't share the same deadline. Hence, they treat the ones with the closest deadlines as urgent before moving on to the others. This makes all members focused on making the best use of their time and ensures that everyone's work matches the organization's objectives.

They Value Themselves as Teammates

Mutual trust and respect are key factors in ensuring the peaceful coexistence of high-performing team members. These two factors go a long way in determining how cooperative the members will be.

High-performing team members utilize this fact and channel it into valuing themselves. They see themselves as teammates, and they trust one another to get the job done. They also respect each other's views when having different ideas. This helps them take risks together, exchange ideas, and follow up on one another's innovative ideas. Thus, they encourage diversity of ideas and see it as the beauty of being a team.

They Celebrate Achievements Together and Recognize Commitments

Since they take risks and accept losses, why shouldn't they celebrate wins together? I know you're thinking the same thing.

They also appreciate every member for their contributions toward the achievements and the organization's goals. Studies have shown that recognition can be shown with more than monetary gifts. Appreciation can come in awards, official gifts, and lots more. This strengthens the bond among the team and ensures a solid culture of joint effort.

They Maintain Genuine Relationships

High-performing teams maintain real relationships. They exchange work ideas and bond over more than official business. A large percentage of high-performing teams are a group of people who derive pleasure in chatting about topics like sports, hobbies, past experiences, and even family. You might wonder, "How can non-work discourse positively affect the team?"

From an authoritarian view, this is time wasted. But the real truth is, it's from the unofficial, personal chats that solid interpersonal relationships are created, and that's necessary for a high-performing team. Talking about issues other than work creates an atmosphere of trust. Of course, you wouldn't tell someone you don't like about your hobbies!

In other words, high-performing teams aren't necessarily the most hardworking, but their unified team spirit leads to productive action.

They Engage in Nonstop Learning

No team has reached the peak of it all, regardless of their achievements. High-performing teams learn from feedback and past mistakes. They search for chances to grow by maintaining open policies, which allows them to accept criticisms in good spirits.

Then they see those feedback as useful tools in becoming better teams. Continuous learning improves teams and keeps them aiming at higher accomplishments.

Factors Posing Threats to High-performing Teams

Even as interesting as these characteristics are, some factors can cause problems for high-performing teams. They include:

- Poor communication techniques
- Lack of mutual respect

- Zero interest in diversity
- Unclear team objectives
- Poor conflict resolution methods
- Negative atmosphere
- Zero tolerance among members, etc.

How to Build a High-performing Team

Consider your organizational layout. What are your strengths? What are your weaknesses? Identifying these will help you know the purpose of building a high-performing team.

Below are some ways by which you can strategically build a high-performing team.

Focus on Communication

To build a high-performing team, one of the key attributes you'll have to incorporate is effective communication. It paves the way for high-performing teams to meet targets. For instance, assuming there is a task to be done, a leader can just tell the team and request their feedback. Here, communication is achieved.

However, effective communication can be hindered by unclear messages from sender to receiver, interruptions, and the inability to decipher the information. But provided that you can find your way around communication hindrances, you will be well headed to effectively building a high-performing team.

Set SMART Targets

There's no team without a target, a goal, or something that drives them. So start by being clear about your target. That's to ensure that the group knows what they are working for. You can also consult team members to get them involved and more dedicated to accomplishing the team's objectives. Make your goals Specific, Measurable, Attainable, Relevant, and Timely.

Since you know that your goal has to be S.M.A.R.T, here's how it should look:

Specific: Are your goals specific enough to challenge your team and achieve your desired result?

Which of these two illustrations is likely to give you the result you seek?

"Let's finish this project and hit the $50k mark by December!"

Or

"Let's work hand-in-hand and achieve $30K in new business from the remote sales team in Australia by the end of this year's first fiscal quarter. Through leads gained from targeted ads on social media and referrals from our staff here working the CRM, we'll hit another $20K. We should have a friendly competition between the team in Australia and the team here; the team that reaches the goal first will have a $5k bonus shared among them."

Now, over to you. Which of the goals above looks more specific? While the first goal is vague and states a desire without a direction, the second goal gives more detail on how the goal should be achieved.

Measurable: Your goal needs to be measurable to help you know if you are headed in the right direction. What does success look like to you? Ensure that your goal has a clearly-defined point of success so that when you get there, you can confidently say, "We've accomplished the task." Once you know your end goal, there should be metrics put in place to help you measure success.

Attainable: Do you have the needed resources to achieve the set goal? Are your teams qualified? List what it will take to achieve success. If you notice that it isn't practical, modify the goal to ensure that it is attainable. For example, if you need to bring more talent to ensure the success of the goal, then you should do that.

Relevant: Does the set goal fit into the bigger picture of the organization? The goal needs to be relevant to the organization's mission to ensure the team stays grounded in their duties and works toward realizing it.

Timely: Once you have a goal, split it into smaller milestones and attach deadlines. This way, team members won't be overwhelmed with an overambitious goal. Instead, they will enjoy the sweet feeling of accomplishment as they achieve each piece.

Prepare Ahead for Conflicts

No good team exists without conflicts. Yes, there'll be misunderstandings at some points. But what keeps a high-performing team at the high spot is the ability to address conflicts. When trying to build a high-performing team, prepare ahead for conflicts. That way, when an issue arises, you'll have measures in place to address the problem.

Envisage the Future

To build a high-performing team, you need to be able to see your team in the future. Be a visionary! Consider your current standings and what you want your team to be in the future, and WORK toward it!

Work on Your Emotional Intelligence

When building a high-performing team, you need to work on your emotional intelligence. A crucial part of this is your emotional intelligence (EI). Your EI is your ability to align your own thoughts and feelings in a way that you can still attend to the thoughts and feelings of others.

To have an excellent EI, you need to be insightful, mindful, and ready to adjust your reactions to blend in the scope of situations.

Coaching

The ability to impact the lives of others is a great art to master. Just as leaders are key players in ensuring that a business yields more output, coaching skills are paramount to helping your team members develop.

Coaching helps your people become critical thinkers, excellent role players, job specialists, and networking agents. On top of these possibilities, coaching others is a powerful strategy for supporting others and enhancing learning.

As opposed to being authoritative and attempting to force people to change, coaching is integral to carrying the whole team along. Research has shown that employees prefer leaders who put them through necessary processes to those that order them around. This type of leader handles (coaches) a skilled workforce, which can help improve business.

There are different categories of teammates that you can coach as a leader. They include:

Category 1: Amateurs

Amateurs are the "*babies*" of the organization. They need the most guidance and supervision. You have to carefully walk them through the nooks and crannies of organizational conduct. And if you're lucky enough, you might be working with a fast learner, and they might not tarry in this stage for long.

Category 2: Practice Crew

After the amateur stage, the next stage is the practice crew. Here, amateurs are starting to absorb the information being passed along. They begin to attempt work at this stage, but they haven't really mastered the skills. But the bright side is that they are contributing positively to the team, to some extent.

Category 3: Performers

As the practice crew move on and begin to get assignments done satisfactorily, they become performers. They attempt real work and put their best into it at this stage.

Also, they're doing the assignments in the manner they should be done. Even though they require less supervision, they still can't be left alone.

Category 4: Professionals

With due diligence and commitment, performers can move up to the professional stage. Here, they can steadily take up assignments and accomplish tasks excellently. At this stage, they've grown strong wings and can now fly with ease. Since they're versed in attending to jobs, they can mentor and educate others.

Category 5: Top Players

You can call top players the specialists, the experts, or the grandmasters. They're key players in teams, and they may rise quickly to become team leaders. They don't create doubts when given tasks because they're exceptionally independent. Like professionals, they can guide other people.

Tips for Successful Coaching

The following are tips to successfully coach any category of employee to become better and more important team players:

Inquire More

Readiness to learn and genuine inquiries lead to more definite and smart responses, which lead to more useful training discussions. It is important to foster solid associations with your members as a team leader or manager. You can consult teammates to determine if they're interested and ready to work toward the team's goal.

Identify What Works Best

To be an amazing coach, you need to be a critical thinker and an attentive leader. Don't try to engage the team in discussions that bore them. Study the team. Do they like recognition? How do they like it? What drives them? You can embark on personal interviews to know what works best for every team member. Endeavor to invest your time finding out what works out for yourself and the team.

Engage Your Team

As a coach, you have to be as friendly as possible with your team. This encourages members to open up to you on certain things about work and even beyond.

Don't excuse members when they come to you. They won't approach you if they prefer to be by themselves. Instead, be an accommodating coach to boost the team's productivity.

Consider Their Viewpoints

Generally, people feel motivated to do more when they see their leaders doing things according to their suggestions. This implies that, while working as a coach, you should make room for contributions from team members. This will inspire them to source more valuable information, deliver them to you, and generate positive outcomes for the team.

Talk About Plans

You should note that coaches engage members in discussions that bring about bigger positive changes. So be sure to point out the next course of action. This will show that you're not just a leader that encourages but also a leader with goals. It'll ensure that you and your team members agree with what you want to achieve.

Be Approachable

Try to respond to your team members when they approach you, even at your most scheduled moments at work. If someone comes to you with a concern, attend to them. If you're too consumed by your work at hand, schedule a later date to discuss it.

You can book an appointment with each team member on convenient days to ease things. This way, you'll be responsive to every member's complaints. You'll also be supervising their progress simultaneously.

Be an Example

No one is an island! Even as you take others through learning, never stop learning yourself. Improve your skills because, whether you like it or not, your employees are watching you, even if it is not obvious. Show them that you're worthy of being followed.

Development

You can just wake up one day, have a cup of coffee, go to work, admire your employees' energy, and decide to enlarge their coasts. While it shouldn't be an "at-your-own-pleasure" thing, investing in employee development has many more benefits than you can imagine. Above all, it's based on the real essence of a true leader: having an impact on others' lives.

Do you need tips on how to develop others? Well, here they are!

- **Be the Force:** To develop others, you must work on yourself. This will prove your worth as a leader. Remember, only diamonds can cut diamonds.
- **Be Open:** Encourage your members by showing them the real you. Don't get them to open up when you're not ready to reciprocate.

- **Put Them in Your Agenda:** Whether personal or organizational, try to include employee development into your plans.
- **Ask Questions:** Instead of being bossy, engage your employees with sensible questions.
- **Share the Work:** Assigning responsibilities to employees helps train them and relieves you of stress.
- **Test Their Research Skills:** Giving employees tasks that are sometimes not their field helps develop them. After going through their tasks, you can know who can diversify and who can't.
- **Create Healthy Connections:** You can introduce your employees to top-notch mentors, role models, or important stakeholders. This will widen their scope.
- **Feedback:** Comment regularly on employees' conduct, achievements, and shortcomings. They'll get better in the long run.
- **Show Them Organizational Ethics:** Besides making necessary connections, take your employees

through what it means to be "for the culture" in organizations.
- **Spend Real Money:** You're the boss, and it's your duty to secure camps, programs, seminars, webinars, and the likes for your employees, even if it costs you real money. Their growth is your business' future.

CHAPTER SIX: BUILD CONFIDENCE IN YOURSELF AND HELP YOUR TEAM BECOME A HIGH-PERFORMING TEAM

"A true leader has the confidence to start alone, the courage to make tough decisions, and the compassion to listen to the needs of others. He does not set out to be a leader, but becomes one by the equality of his actions and the integrity of his intent"

- Douglas Macarthur

Confidence has to do with the feeling of certainty within you. You can call it trusting one's ability. It's a state of mind where you control all your emotions. At that stage, you're the master of your emotions, especially fear. Self-confidence isn't always precise. Your confidence to attempt some things or perform tasks and deal with circumstances can either increase or decrease on any given day.

On some days, you may feel more confident than others. Being confident doesn't necessarily mean that you're "perfect" in everything. No! You'll make mistakes. Mistakes are inevitable, especially when doing something you are not used to or something you have not done before. But confidence covers your ability to learn from your mistakes, grow wiser, and move on to the next lines of action without allowing your mistakes to drag you down.

Building Confidence in Yourself as a Leader

Leadership grows based on self-confidence. Trying to show leadership without developing confidence is like erecting a building on a sandy foundation. You may coat it with nice paint and make it look nice, but you've just erected a death trap. That's it!

Fear has seized the minds of many leaders, and that has caused drastic effects in many organizations today. Major business leaders focus on the leadership skills of communication, empowerment, and passion but ignore the importance of confidence. You should note that without building around confidence, you're likely to face hardships in being an effective leader.

Here are tips to help you be more confident in yourself as a leader.

Silence Your Fear: Be Aware That Your Fear Is Also Someone Else's

Fear is the major hindrance to being confident. You have to overcome it before it wrecks great havoc on you, your team, and your business. One way you can accomplish this is by having a role model.

We all need role models, whether we like it or not. Having role models helps you realize that you're not the only one that's afraid. It'll show you that your fear has been faced by someone else and has been subdued. You don't need role models just at the early stages of your life; we also need them throughout our endeavors.

Like Sandjal Brugmann, founder of The Passion Institute, said, *"Don't struggle with your fear—learn to embrace it and stop preventing you from following your dreams."* You need to let go of the unending thoughts of fear within you. When you constantly think of fear, you'll be tempted to hide or run from it, rather than seeing it as something you can chase or control.

Developing your confidence strengthens you against all types of scary feelings that you might experience. Being aware that you're not the only one in your situation will help you greatly by giving you a sense of belonging. It'll assist you in seeking help outside because you know that others are or have been in your shoes.

Fear endangers you, your team, and the organization, causing you to make decisions that are not in line with the set goals. In other words, if you allow fear to be in control, you're at the risk of damaging your own goals. Hence, it's a red flag for entrepreneurs. You can be scared to invest a high capital into a new venture or of having to communicate with various personalities—team members or clients. Knowing that your decisions will not only affect you but also affect your team and the whole organization can also instill fear into you. No matter what it is, try to subdue that fearful thought by all means.

Pick a Role Model

A role model is a person others see or look to as an example of what they want to be like. A role model is worthy of imitation. While a model represents an inspirational idea, a role model inspires others to embrace their characteristics. Research shows that your confidence increases when you find out that your heroes have succeeded in facing similar situations as those you're currently passing through. You get to know that it's possible to make your dreams achievable, no matter how big the circumstances you face may seem.

Why you're not yet confident might be due to either not having a role model at all or having the wrong role model. In both cases, you're not gaining any good. Since it's natural to compare yourself with someone in this competitive world, find a role model you can compare yourself with. In fact, many leaders believe in healthy comparison.

The Right Role Model

When talking about appearance, composure, and career life, what the right role model portrays is what you need. Hence, the right role model for you is similar to you. They can provide you with inspiration and information about what you're facing and, at the same time, the needed confidence to overcome that which you're facing.

Build Profitable Connections: Networking

Building connections with other leaders and businesses is a trusted way to share real-world experiences and find solutions. Leaders are aware of the importance of networking, and many have keyed into it.

Networking is not only meant to be done at the initial stages of your career but throughout your days as a leader. Solid connections or professional networking helps boost your confidence. You will meet people who admire your work and people you will look up to. It could be your team's pathway to becoming a high-performing one; who knows?

Here are some tips that can assist you on how to build solid and professional connections:

- Begin your networking from within your team or organization.
- Join professional bodies.
- Attend related conferences and seminars.
- Try mentoring.

Control What You Feel

Personal views matter. What you feel within yourself can either be of doubt or of confidence. And it can have significant impacts on major areas of your life, especially your career. Now let's get into the two important personal feelings.

Self-confidence

Self-confidence is a positive view about yourself where you see yourself capable of doing anything. It's important that you develop this feeling as it contributes positively to your life, your team's spirit, and your organizational goals.

Self-doubt

Being uncertain about yourself and not being optimistic is basically self-doubt. It's a negative feeling about yourself where you think you're not the right person for the task. Try not to allow it to control you into making really drastic decisions as the consequences might be unbearable.

Acknowledge Your Strengths and Work on Your Weaknesses

Your well-developed traits are your strengths. They can be your leadership skills, communication, management, the delegation of tasks, etc. They're considered positive. Weaknesses are just the opposite. While it's improper to term weaknesses as "negative," it's necessary for you to work on them. Weaknesses include personal fears, team shortcomings, and poorly developed skills.

You should note that having a weakness doesn't mean that you're bad or have lost it all in that aspect. It means that you have to work on that weakness because it's a strength that's not as dominant as the others—and you want it to be.

A trusted way of overcoming your weakness is knowing yourself. Many people don't know themselves, which draws them back from maximizing their potential. Knowing yourself fully helps you feel eased and settled in attending to your affairs. What may seem like strengths might not be the required strengths, but when you know yourself, you'll easily identify your real strengths disguised as a weakness.

Be Mindful of What You Consider as Your Priority

As a leader, you should know the differences between "daily routines" and "priorities." You can just wake up and attend to some tasks that are "worth your time." That doesn't make them priorities;

they're daily routines or chores you're expected to do. Sometimes, your "priorities" can be to continue a task from the day before. That shouldn't be!

You have to carefully search for priorities so that you can channel your energy into attending to them—whether you're confident about accomplishing them or not.

Here are some ways to choose priorities:

- Ensure that you're the one choosing tasks and not the other way around.
- Decide which are important and urgent tasks.
- Choose the takes that are related to your aims.
- If you're still uncertain about your priorities, you can choose the tasks that will make way for other tasks.
- Pick the tasks that make you feel uncomfortable.

Try New Things

You've probably been doing the same thing over and over again. While this helps you feel safe, it doesn't build your confidence because you're already used to that activity.

Trying new things outside the scope of what you've done might put you, your team, and your organization at risk. However, it paves ways for new results to surface, thereby increasing your experience level and your confidence.

Develop Your Record-keeping Skills

Record-keeping means that you regularly check what you've done and take records. This will help in having references to fall back on before making decisions. Your confidence is boosted when you see how far you've gone personally or with your team.

Be Kind to Yourself

Being kind to yourself is appreciating yourself. You can start treating yourself differently and be kind to yourself. This could be in the form of keeping your cool and talking to yourself or reminding yourself of your accomplishments. Don't argue with your inner critic; you'll feel worse than before. Instead, you can repeat words that boost your confidence, either aloud or in your head.

Practice Self-Appraisal with Humility

Speaking to people about your strengths reflects and boosts confidence, although many don't know how it boosts confidence. Confidence is knowing what to do and doing it effectively without feeling uncertain about your actions. Speaking about a specific strength that shows your level of experience at handling a particular task boosts your confidence and helps in motivating you to accomplish greater tasks. Strengths that you can talk about include:

- Character-based strengths like honesty, trustworthiness, integrity, and loyalty
- Skill-based strengths like management, record-keeping, and customer services

Trust Your Gut

You know all about the five human senses, right? *To burst your bubble*, there's a sixth sense—the gut feeling.

No matter what you choose to call it, it provides you with the motivation you need. At times, you might be stuck in a situation, but you suddenly picked up insight on how to go about it—that's your gut!

Trusting your gut means that you're following directions from within yourself, and at the same time, you're being honest with yourself. Most times, the results are positive. Thus, it boosts your confidence.

However, following your instincts might be risky, especially when you haven't done that before. You might be confused about trusting a feeling or an instinct you can't explain. But you should note that doing something your own way gives you the confidence you need. You'll feel fulfilled and be able to bear consequences without fear of uncertainties or anything else. In short, trust your gut, do your thing, and don't be a copycat.

How to Instill the Above Into Your Team

Firstly, you need to work on yourself as a leader. After that, you can dive into building confidence in your team. You can build your self-confidence by:

- **Reminiscing on previous achievements:** This will give you the happiness you need to free your mind and believe more in yourself and your abilities.
- **Thinking of things you're good at:** This will help you find more ways to be better at those things and develop other skills as desired.
- **Setting goals for yourself:** It helps in strengthening your ability to focus. The more focused you are on a goal, the more confident you become in achieving it.
- **Talking to yourself:** Ridiculous as it may sound, it helps in being aware of yourself.
- **Getting a hobby:** Hobbies like singing, dancing, reading, and exercising have positive health effects. They help in building confidence as a result of the accomplishment of tasks. For

instance, you might have been a 100m sprinter; once you attempt 200m and you're good, your self-confidence will be on the rise!

Provided that you've succeeded in developing your self-confidence, your team mustn't be left behind. You should carry them along in the journey.

Building Confidence in Your Team

As a leader, you ought not to be the only confident one in the team. Your members should also feel confident. When your team members aren't confident, they won't be able to perform at their best. It's therefore imperative that you build confidence in your team. Should you notice that a member is suffering from self-doubt or lack of confidence, you should:

Offer Help with Learning and Development

While learning can negatively impact people, self-doubt, confidence, and professionalism are very much connected. Learning new things can leave you with the feeling of "Can I?"

which drains the energy to be productive. Hence, as a leader, it's expected that you help your team members learn new things. Be with them throughout the learning process. This way, you're not only showing them new skills, but you're also motivating them to be confident in themselves the same way you are.

Then, as time goes by, encourage them to put what they've learned to practice. Doing what they can do, especially when they have the experience, can build their confidence because they will feel assured that they can do it.

At work, organize training sessions for employees. They should be taken through jobs, processes on getting the jobs done, updates, and workplace conduct. And if there are changes to be implemented, don't hesitate to teach them ways by which they can handle those changes. Do this because sometimes, people become destabilized when changes occur.

Assign Step-by-Step Tasks

Assigning tasks to team members involves sharing tasks to achieve the collective goal. Here, the delegation of duties depends solely on your judgment. Every member of the team is expected to cooperate in performing their duties. To successfully assign tasks to members, you must first take care of the difficult members because they'll always undermine your authority. When everyone is on the same page, the delegation of duties becomes more interesting and easier.

Delegation of tasks can:

- Save time which can be used for other profitable activities.
- Create a balance between serious teamwork and extra-work activities.
- Create a strong relationship between you and your team since they have to report back to you.
- Speed up the team's work process.
- Boost members' confidence in performing their tasks excellently.
- Ensure that members are prepared for advancements in their professional careers.

- Show members that you're a performing leader.
- Make team members see you as an expert that they can approach for clarity on some issues.

Focus on People's Strengths

You need to stay connected to your team if you're serious about building your team's confidence. And a good way to do that is by focusing on every member's strength. Although you'll have to help them work on their weaknesses, show them that their strengths are what the organization needs.

This will boost their confidence and lead to massive development, better customer services, excellent work performance, fewer conflicts, smooth management, and greater productivity.

Show Support

We all know that humans are social creatures. Our ability to come together, get organized, and work hand in hand to achieve a common goal, no matter the level of difficulty, is top-notch. That's why we're still the most celebrated species.

As expected of a leader, support every aspect of the life of every member of the team. Being supportive involves backing them up, giving them the go-ahead, and calling them to order when necessary. This will boost their confidence since they will see that their leader believes so much in them.

Showing support also makes you an example worthy of emulation. It'll encourage employees to support themselves at every point in time. Their support can also generate confidence in the team since everybody's got everyone else's back.

And to be supportive, you need to put yourself in their shoes. Feel what they feel and help them in those areas where you know they need help. It can be work or non-work areas; it doesn't matter. How you treat your members will, in general, give them a lasting impression of the type of team they belong to. Support them through your way of addressing them, the delegation of tasks, and the way you take them through processes in the working environment. This is because your relationship with them will determine if they'll be inspired to perform or not.

Research has shown that team members who feel that their leaders think of them with high esteem work better and vice versa. Both scenarios are evident in today's organizations around the world.

Embrace Failure

Failure shouldn't be seen as the end of it all. You should make your team see that failure to achieve a team goal doesn't mean the team is ineffective; it simply means that the team needs to work more.

Although every one of us knows that failure isn't an end to everything, we still find it difficult to embrace it. Instead, we get depressed, weak, and unbalanced. You and your team need to be able to work around failure.

Naturally, failure disrupts a team by causing conflicts among members, bad management, lackadaisical attitude at work, and personal and professional dissatisfaction. However, a great team is judged not only by their numerous achievements but also by how they handle shortcomings, failures, and disappointments.

Embracing failure isn't as easy as it sounds. It requires knowing why it's important to embrace it and how you address your team, should it occur.

Why is it important to embrace failure? The most important reason is that not embracing it will deprive the team of more opportunities. It will make all team members lose morale, energy, and focus in putting more effort into work. It can also bring about the team's dissolution.

Regular Feedback

You shouldn't underestimate the impact of feedback or responses in a team. Whenever you decide to assign tasks to team members, don't forget to see yourself as the director. You need to encourage your team members to always give you feedback on tasks assigned to them, decisions you make, and discussions at team meetings.

Your team members are a group that you should carry along on issues pertaining to the team's goal. Be responsive. Tell them how well they did their jobs, appreciate them, and ensure that their efforts are recognized. Don't be caught in the web of "giving responses takes too much time, and I'm even uncertain about what to say." While it's important that you let your members know how you feel about their performances, ensure that you employ constructive criticism methods to avoid hurting any member's feelings and lowering the team's confidence level.

This way, you'll know every member's standings as far as their performances are concerned and proffer solutions if there are any lapses.

Celebrate Achievements Together

As a leader, whenever you're celebrating achievements, show your gratitude by recognizing individuals who assisted with getting it together. Don't ascribe the whole glory to yourself. Try to help others in succeeding so that they can feel inspired to celebrate as well.

CHAPTER SEVEN: EFFECTIVE COMMUNICATION SKILLS AND MOST PROVEN TECHNIQUES

"To effectively communicate, we must realize that we are all different in the way we perceive the world and use this understanding as a guide to our communication with others."

- Tony Robbins

Effective communication is one of the best skills you can possess as a leader. It helps you send information to others and interpret what is said or sent to you. The best scenario of effective communication can be seen between babies and their moms. They listen eagerly to their mom and attempt to decode the sounds that she makes—get what she's trying to pass across. From that, you'll see that the desire to communicate is inborn.

Communication, also called correspondence, sends information from one place, person, or spot to another. It can be through the use of voice (verbal), composed media (like books, sites, etc.), designs (pictures, logos, etc.), and signs or signals (non-verbal). You've used a blend of a few of these in the past.

The Importance of Good Communication Skills

The main importance of good communication skills is to build your relational abilities. Strong relational abilities can help in all areas of your life, from your career life to informal relationships.

Passing information precisely, clearly, and as expected is an indispensable fundamental ability and should not be disregarded. And if you've not yet keyed into this ability, it's not too late to deal with your relational abilities. This way, you may see changes in your personal satisfaction as your relational abilities grow.

You need to pursue greatness, promotions, and positions in your career. Being a good communicator will help you achieve these. For example, to address a wide gathering, you need to use the right words and expressions related to that gathering. You know that you're expected to create your address to suit that gathering. That's basically a relational ability, and that's what you need as a leader. As you move on, you'll see the importance of effective communication: the ability to talk appropriately, blend in discussions, receive information without misinterpreting it, and be composed.

Effective management requires effective communication. Poor communication has been the reason behind the dissolution of many teams, organizations, and networks. Being able to interact effectively also requires the ability to listen.

Communication skills can likewise help in ensuring smooth management within teams and organizations. Moreover, these skills are paramount when considering ways to achieve progress in all facets of life. Thus it's beneficial to figure out ways to improve your communication.

How to Improve Your Communication Skills

To be a good communicator, you need to possess certain skills. They are essential to ensure that all parties involved in the communication process are satisfied. By accurately employing these skills, you can prevent conflicts, misunderstandings, or misinterpretation of messages.

Also, these skills will greatly help you in reaching a close-to-perfection level of effective communication. They include:

Practicing Active Listening

One important skill that you must learn is active listening. You can practice this by constantly and consciously training. Although paying absolute attention might be hard because distractions are never-ending, you'll master it eventually.

Being an active listener implies that you're all in on that discussion, chat, or idea exchange. It means that you're focused on what is said rather than just randomly "hearing." Active listening helps you and the other party to decode messages faster. You can read what your colleague is saying even from their gestures and body language. They will also read your reactions and see if you're interested in the conversation.

An active listener shows interest in who or what they are listening to. But how do you show interest? You can do this through two media which are:

- **Verbal Media:** This involves replying to the speaker to show that you are on the same page. This method involves saying words like "yes" and "okay" where appropriate, as well as replying to questions thrown at you.
- **Non-verbal Media:** This method involves using nods, smiles, and eye

contact to tell the speaker that they can proceed.

Making good use of these media will help give confidence to the speaker, hence communicating clearly and effectively.

Being Concrete and Clear

When addressing your team members, be clear about your message. Why are you speaking with this person? What's the message you're trying to send to them? You need to be sure. If you're not sure of your message, your audience won't be sure either.

Also, in writing, try to reduce the number of thoughts in a sentence. Make sure that what you're writing is simple enough for anyone to understand. Not everyone derives joy in finding meaning in what you write. Some don't like it, and some find it time-consuming.

Making yourself clear helps your recipient absorb your messages instantly, rather than having to ponder what they mean. Clarity has the following benefits:

- It makes understanding simpler.

- It brings out the essence of the message.
- It gives room for the use of suitable and related words.

When your message is concrete and clear, your audience will have no problem getting the full picture of what is being communicated.

Concrete communication infers being specific and clear rather than fluffy and general. Being concrete backs up the validity of the message. Just like anything, a concrete message has features or characteristics. Some of which include:

- A unique viewpoint
- Clear words with clear meanings
- Not likely to be misunderstood

Non-verbal Communication

Most times, we see communication as just what we say. That's not correct. Effective communication goes beyond and has more meaning than the audible messages you pass across to people.

This is where non-verbal communication comes in.

Non-verbal communication covers the aspects of looks, gestures, or motions shown (kinesics) and the amount of separation between you and your recipient (proxemics).

Non-verbal communication skills can provide more important information than verbal skills. Research has shown that over 70 percent of communication is done through non-verbal means. Non-verbal interaction isn't a skill with a static benefit. However, it's influenced by the present situation. This covers the location (fun joint, office, home, etc.) and the individuals involved in the process.

For instance, a head gesture between partners in an advisory group meeting might mean that they're agreeing on some terms, while the same gesture of the head can also be used to acknowledge the recognition of someone in a jam-packed room instead of shouting at each other.

Non-verbal communication is comprised of body language and sign language, which should be deciphered alongside the words.

Banking on Relaxation

Everybody encounters different anxiety levels, and anxiety can result in a lack of boldness or being nervous when working. Unfortunately, there's no specific location for totally eliminating anxiety because even at the most relaxed locations, some people won't feel relaxed. So try to figure out how to relax.

While there are different ways to relax, the most discussed ways are:

- *The "Hand on Hips and Separated Legs" Method:* As the name implies, you have to place your hands on your hips and keep your legs wide apart. Although this method looks like you're trying to practice meditation, it assists individuals in building confidence.

- *Deep Breathing Techniques:* This basically helps you control your breathing in case you're short of air due to anxiety. Thus, it keeps you relaxed.

Make sure that you're relaxed as much as possible. The more relaxed you are, the lower your anxiety level. Hence, the more confidence you and your team will get.

As a leader, ensure that you act like yourself when communicating with your team members or business associates. Transparency demands that you communicate with people in all honesty— not being fake about what or who you are. Only then can you lead with a relaxed mind because you have nothing to hide. It's only a relaxed leader that can successfully lead team members.

Striving to Inform

Today, many information communication centers equip people with the necessary knowledge, skills, and processes more effectively. Why? Because the main goal is to inform!

Creating informative messages is a good way of building good communication skills. These informative messages include:

- Short, straightforward messages
- Long, formal answers
- Professional introductions
- Project plans

These are evident in work guidelines, processes and systems, work updates, work results, and reports. Since the reason for informative communication is to enlighten others, adopting this type of communication will help your audience interpret what the message being passed across really means through clear words, thorough explanations, and illustrations.

Informative communication isn't planned to adjust your audience's perspective on an issue, even though that might be an accidental outcome. Therefore, whenever your motive is to enlighten, ensure that you present your message clearly, are easy to understand, and are focused on delivering the right meaning to the message.

Making Use of Visual Communication

Here, the goal is to convey your thoughts to others through symbols and images. Visual communication is the third type of communication aside from verbal and non-verbal, and it's seen as the most important to people. The reason is that everyone takes note of pictures, symbols, moving pictures (animations), videos, and different models—no matter how small the details may be.

Let's assume that you traveled to a new city for the summer holidays, and you got lost while driving. You know that, right then, there's no one to call on, and the city is not a familiar one. Perhaps you can't speak their language; you just know that asking for directions won't work out. So then, you know your last resort is to use everything around you, like signposts, natural signs, tourist spots, etc., to successfully lead you back to where you live. Here, we can say that you've used visual communication in retracing your steps back to your vacation lodge.

Showing Empathy

Like every aspect, leaders need to show empathy when communicating with their team members.

Empathy is considered immature in business; this is why it's the communication skill with the least usage. However, it's useful in effective communication as it builds trust, settles disputes, and conveys ideas.

Empathy isn't only understanding. It simply means that you can view an issue from another person's (maybe a team member's) perspective to understand their experiences and contributions. It doesn't mean that you comprehend. It doesn't mean you're in total agreement. It basically means that you're trying to make a person see their view as being understood while trying to understand their views personally.

Completeness

Whatever you're trying to pass across to another person must be complete. Completeness demands that you consider the recipient before communicating. This act of "getting everything covered" helps your recipients see the real message and how it affects them. In addition, sending detailed or complete information implies that the communication process is successful as all parties involved are being considered.

Apart from both parties' consideration, a complete message answers all the six W/H-questions: *who, what, when, where, why, and how.* Of course, this doesn't mean you should address those questions one after the other whenever you want to send a message. No! It simply means that you need to include the appropriate responses for your message to be a complete one.

And in case you want to create a complete message, here are some highlights on what a complete message does and what to look out for.

Complete communication has the following highlights:

- It makes the standing of an organization known.

- It saves cost because no information is absent, and no extra cost is used in getting additional information.

- It supplies extra information where it's required. That is, it leaves the recipient in clarity of the message delivered.

- It helps the recipient make quick and better decisions since all required information has been put across.

- It convinces the recipient.

Giving Feedback

Another important communication skill is giving feedback. However, this skill takes practice. That's why it's important to know how to give useful feedback. Leaders and members respect the act of giving feedback that it has now become standardized. Business owners now request weekly, month, and yearly reports from employees. This is because it helps ensure that the business owners communicate effectively with the employees.

It's unsurprising that giving feedback can be scary. The reasons range from self-doubts to uncertainties about the recipient's reactions. And these reasons contribute to an unhealthy atmosphere that doesn't support effective communication.

You should note that giving and receiving feedback is critical to connecting with people around you, including your team members, and keeping them on target. For instance, you need to motivate your employees to do more; you can let them know how good their performance is. This will stir them to execute their tasks more exceptionally. However, be genuine and moderate when giving feedback.

Celebrate Success

Celebrating success is essential to effective communication, especially in our world today. Effective communication ensures that information passed across is received. Thus, it's important to try informal means in achieving it. That's why celebrating success is important.

Celebrating success creates an avenue for people to interact among themselves. There's a high possibility that serious matters can be communicated. For instance, during the celebration of a team's achievement, you can give a congratulatory speech as a leader. You'll emphasize the merriment and acknowledgment of efforts and chip in necessary information regarding another achievement that's yet to be attained.

You may be wondering, "Why celebrate success?"

Well, the reasons aren't far-fetched, and they are:

It Motivates You to Do More

It's natural for us to feel more motivated when our successes are celebrated. It gives the feeling of "I need to do more," thereby making you aim higher. At work, success can be celebrated through promotions, recognitions, or increment of salaries, and these energize employees toward achieving more for themselves and the organization.

Being successful gives individuals the idea that they're capable of accomplishing tasks. In addition, celebrating success gives the impression that they can always repeat these accomplishments. Celebrating success is crucial because it builds the confidence needed for effective communication.

It Shows How Far You've Come

Celebrating success helps you look back and imagine where you started. It brings back memories of when you've achieved other things, making you feel alive.

Review your set goals. Have you been able to achieve them? If yes, the next line of action is to set new ones to achieve more. If not, a lot needs to be done. In other words, celebrating success allows you to brood over the past, see what has been done, what you're doing, and what you should do. Then you'll see that every effort is worth celebrating at the end of the day.

Doing this will fuel your confidence and help you progress into getting the best out of yourself, your team, and your organization at large. As a result, there is generally more than you can accomplish, more abilities that you can create, and more progress that you can make.

Studies reveal that looking back at the past gives its own feeling of accomplishment as everyone loves to see that they have made a difference over time.

Thinking back and celebrating your achievements can make you more determined against all odds because you already know that every challenge prepares you for what's ahead.

Improving Effective Communication

Learning and improving good communication skills requires regular practice, commitment, and conscious effort. You can also take courses, research about effective communication, and practice. However, don't ignore the importance of role models, mentors, and experts. They can help you alleviate your fears and develop better communication skills to the best of your ability.

The following are ways by which you can improve effective communication:

- Engaging active listeners. Active listeners are great. Don't forget to practice active listening.
- Adopting non-verbal communication techniques. This will assist you in assessing your audience, checking for interests, and avoiding boring communication.
- Appropriate expression of feelings
- Research on ways, techniques, and methods known to improve your relational abilities
- Don't shy away from public speaking. It'll amplify your skills, improve your weaknesses, and develop your effective communication skills.

- Create networks. Having connections with other people will allow you to gain from their experiences, learn from their mistakes, and bank on their successes. This will guarantee a safe path to effective communication—anywhere, anytime.

CHAPTER EIGHT: NEGOTIATION TACTICS AND DEALING WITH DIFFICULT PEOPLE FOR A SUCCESSFUL OUTCOME

"Successful negotiation is not about getting to 'yes.' It's about mastering 'no' and understanding what the path to an agreement is."

- Christopher Voss

Negotiation is a popular word for a meeting between two or more parties to find common ground. It's an interaction and process between entities who aspire to agree on matters of mutual interest while making the best use of their individual services. Also, it's how differences or conflicts can be resolved by reaching agreements or compromises. In other words, for a negotiation to take place, there must be an issue or a conflict to deal with.

In a negotiation, one party tries to win the other party over to its side. This process involves comparing viewpoints, finding faults in each other's view, pointing out the benefits of both views, and reaching a conclusion or an agreement on the single path that both parties have to follow.

Why Is Negotiation Important in Leadership?

Negotiation helps settle conflicts about ideas, resolve workplace disputes, or get everyone on a team to be on the same page. The ability to negotiate is important in our personal and professional lives, although it's more important in the latter. This negotiating ability helps in self-management amid an unfavorable situation. And it's very evident in organizations; they expand their reach by bargaining with other organizations.

There is truth in the saying "Great leaders aren't born, but made." This implies that anyone can become a great leader through constant and conscious efforts toward development. One of the most important ways of knowing a great leader is through their negotiating power—the ability to bargain successfully.

You might wonder, "Why should leaders, managers, and business owners negotiate?" The answer isn't far-fetched. It's based on the fact that negotiation is an important way to succeed. For instance, employees negotiate with their managers for payments.

Excellent negotiation isn't an easy skill to master. First, you need to overcome your feelings of doubt, fear, and pessimism and pay more attention to the situation.

When you bargain impressively, you stand a chance to create connections with various individuals, team members, or other leaders because you can strike deals in a way that's acceptable to them.

Also, instead of just waiting on your team and looking at their work, you can gather them around, discuss your expectations, hear theirs, and reach an agreement.

Your negotiating ability assists you in accomplishing your tasks and contributes to your overall success as a leader. And you know that your success affects every part of you, whether as a professional or not. Therefore, as a leader, you need to learn how to bargain so that you'll secure things that will benefit you, your team, and your organization.

The following points are why having good negotiating skills is crucial.

The Leader and Team Members Benefit

Your negotiating power is very useful to everybody, be it a leader or team member. It's a significant skill that, when mastered, can help you start securing profitable deals and having new networks of business associates.

Win-Win Circumstances

Negotiation involves both parties considering and comparing each other's ideas and benefits. Hence, it's a win-win situation if you and your negotiating partner agree to bank on those two-sided benefits. However, this kind of situation comes with its own challenges. Therefore, you'll need to be careful and smart when negotiating.

Works on the Eventual Outcome

Undoubtedly, the main reason behind negotiations is to get the best outcome that will favor you and your team. With good negotiation, you can get the other parties on your side of the fence. This can help create the most ideal arrangements for yourself and your team.

It Makes Your Team Submissive

Good negotiating skills will help you persuade your team members to perform at their highest efficiency level. This also means that they'll respect you and be loyal to you.

The following are strategies to fix a very good agreement between two parties.

Achieve a Win/win

To achieve a happy ending in the negotiating process, you should cautiously investigate your aims and those of your partner. This is to find a commonly acceptable outcome. If you're both happy with what you've gotten at the end of the negotiation, you've achieved a win-win.

To know if it's a win-win situation, recognize your negotiating partner's viewpoint as very useful for you. They should also accept your viewpoint. If this isn't the situation, one of you will have to give in to the other's idea. Then compensation may be given to whoever submits to the other party's idea. However, the two parties must be okay with the outcome—no matter how complicated the negotiation process is.

A win-win outcome is the most desirable. Although it may not generally be imaginable, it should be your aim when negotiating. However, a compromise is your best bet instead of a win-win outcome—which may not be totally possible. It'll help provide a good alternative and leave both parties satisfied.

Sometimes, you might go into negotiation with the aim of being the only satisfied party. It can happen, but it's not always certain. Try to work things out so that you and your partner are satisfied. Your bargaining power is the most important factor in ensuring that you get the best result for yourself and your team.

Be Emotionally Adept

Emotions are built-in and are the components that guarantee everyone's endurance. They serve as proof of solid qualities, interests, and convictions. They give meaning to whatever you do and help you impact or convince your team members to do your will. Emotions help you to learn and hold on to what you've realized. You should be in control of your emotions, since it's never possible to get rid of them. So how can you control your emotions during a negotiation?

Most people think their emotions are like a switch that they can just turn on or off whenever it pleases them. However, that's not the case. Facing your emotions allows you to subdue their effect on you, thereby putting you back in control. Next time, work on yourself instead of trying to run from yourself.

To be an excellent negotiator, you should first analyze your true emotions and know what they are and what stimulates them. Then find a way to deal with them. Only then can you care about the emotions of others.

Are you the careful type? Are you the type that's very mindful of what triggers your words or actions? No matter what your answers are, the point is that you need to be aware of yourself. Awareness helps you decide how to deal with your emotions instead of just responding to them.

Raising your awareness includes:

- Investing your energy into deals that might not turn out successfully.

- Analyzing how you responded when emotions took over you during a negotiating process.

Don't consider what you say alone. Check for your inner reactions. Do you feel compromised, powerless, furious, or restless? How often do your emotions affect you during negotiations? You can seek help by asking mentors and knowing their views on how to control your emotions.

The next time you're in a negotiation, take note of yourself and what you feel. Does your body feel strange when you're stuck with an idea? Do you have shallow breaths, weak muscles, or do you just quiet down? Do you talk more or less? Whatever it is, know that your body will always respond to your emotions. As such, be in control. Use what you figure out to define a few objectives for yourself on how you need to deal with your feelings.

Ask for More Than You Expect to Receive

This technique is very useful when you have a partner that immediately jumps into negotiation. If they request more at first, respond with the same energy by requesting more than they can offer.

Assuming that your partner needs a lower cost, sell them more items. It's simpler to negotiate for more sales if you've allowed the other party to believe they've won the negotiation. According to Ryan Stewman from Break Free Institute, *"Allow them to win the first offer, and they will automatically be happy enough to allow you to win a few more."*

Don't Make the First Offer

This simply means that you don't have to be the first to start a negotiation. There are two types of parties in negotiation—the excited and energetic one and the hesitant one. The hesitant one, in the negotiation, has more command over the negotiating speed and outcome. So to be a very successful negotiator, you should try to be the hesitant party.

There are a couple of things you can do to show that you're the hesitant type in negotiation, and they include:

- Try not to make the first offer.

- Demonstrate hesitance in non-verbal ways such as sitting back in your seat, crossing your arms, etc.
- Talk calmly and serenely.
- Seek clarification on some unclear questions and challenges as the deal goes on.

Assuming you play the role of the hesitant party, you're indirectly forcing your partner to pick the energetic option. This sets you in a situation where you're in control of the negotiation process. And when it's time to make your own contributions, setting the bar higher than what you might want to receive is very useful in bargaining. Your negotiating partner will surely be amazed at how good your skills are when done appropriately. Have a goal and take action toward it. It'll help in falling on the good side during negotiations.

Also, as a negotiator, you should explain your views and their impacts in a bid to trade ideas and convince the other party of your perspective.

The best negotiators are those who pass across information effectively. To do that, you need to study your negotiating partner. Assess their possible responses and prepare your relational abilities to tackle the responses and the unexpected ones. Inability to pass across information effectively will keep you on the low side of the bargain, while maintaining good relational abilities will help you with all you need in negotiation.

Knowing When to Stop

Knowing when to stop is another important negotiating skill that you can build on. Whether you're done with the bargain or not, be sure to stop when you have to. This shows your partner that you are considerate and care about their perspectives. In addition, it makes them voice out their own interests or counter yours.

For instance, if you keep switching from one idea to another without giving your partner the chance to reciprocate, you'll look unprofessional and untrustworthy.

Knowing when to stop shows that you're confident and have a solid initiative.

Advantages of Good Negotiation Skills to Organizations

You can't underestimate the benefits of good negotiation skills. Your negotiating power influences your overall performance, team, and organization.

Here are advantages of having good negotiating skills:

Develops Your Self-Confidence

Negotiation requires certainty, and being certain requires confidence. Anytime you bargain, you're facing an entirely different organization. You are automatically your organization's representative. Thus, negotiation helps build your confidence when you face other organizations.

Allows You to Maximize Value

Maximizing value comes into play when you have good negotiating skills. You can do this by not jumping into the negotiation first. This will allow you to have more chances to make the best out of the issue. Also, you should be aware of the consequences of the bargain so your side can receive the benefits later on. All things considered, it's for your organization's benefit to capitalize on your speculations. When you've fully developed your negotiating skills, you'll have better ways of guaranteeing that you have acquired the best out of the negotiation.

Forestalls Problems

Negotiations tackle problems. Organizations are consistently tested by many problems, from everyday conflicts to professional debates.

Good negotiating skills will significantly empower you to adjust to the many difficulties you face and keep any major issues from emerging. For example, a problem within a team can be forestalled when you've moved into negotiation before doing anything. However, there'll be various issues that might happen unexpectedly. But when you're great at negotiating, you can walk over these obstacles and succeed.

Keeps Stress at Bay

Regardless of your duty in the work environment, there might be a wide range of distressing circumstances you need to deal with. The ability to negotiate is important in every organization. You can haggle with your co-workers and colleagues viewing different work perspectives like assignments, deadlines, and choices. With your good negotiating skills, you can deal with potential problems and stress in the workplace.

Builds Respect

Respect isn't something that can be underestimated in business. For processes to work, respect must be given wherever it's due. For instance, you need your team members to achieve the organization's goals. For them to obey you, they must respect you as their leader.

Good negotiating skills allow team members to see the need to respect you and the organization.

Settles Conflicts

Good negotiating skills help prevent not only conflicts but also settle existing ones. Of course, conflicts can occur due to misunderstandings, but calling both parties to order and reaching a compromise will help settle differences in the organization.

To successfully settle conflicts through negotiation, you should hear each party's interests. Also, you need to evaluate and weigh whether their interests and support are justifiable or not. This will keep your organization moving, building great networks, and being trustworthy.

Works on Your Organization's Reputation

Your organization's reputation speaks for itself, you, and the team. So no matter the situation, ensure you apply straightforward methods in your negotiations. This will help in building your organization's image.

When you have good negotiating skills, your negotiating partners will respect you and the organization you stand for. In addition, they will recognize you as knowledgeable in business information and with good abilities. Your team members and employees will also show you due regard.

How to Deal with Difficult People

Within an organization, all employees and leaders are expected to perform their tasks effectively and get good outcomes so that everyone will be happy. Unfortunately, that's not the case every time because there are always bad apples in every bunch. Some people, intentionally or not, do not perform effectively.

Some people are unproductive compared to others. And if nothing is done, their unproductivity can cause problems and affect the organization. However, dealing with difficult people sets the difference between just being a leader and being a great one.

The following ideas will help you effectively deal with difficult people:

Be Understanding: Know That We Don't Have Equal Capacities

Most leaders often misunderstand how life goes. They think that everyone should do the same because they can perform some tasks. No! We're not equal in ability, intellect, and endurance. That's why as a leader, you should be considerate.

If someone is new to work, they have a high possibility of anxiety, which might come in the form of resistance. They might even find it hard to perform well at first, but that doesn't mean you should rule them out. Don't judge people based on first impressions. Instead, try to guide and train them. After that, you'll surely see changes in them. A great leader will come to other people's level, feel what they feel, and provide them with training, mentorship, and support.

Assign the "Right" Roles

Most people are not in "their" jobs. This is not to say that they're not in desirable jobs, but they're not playing roles that require their full strengths. Being in the wrong job will affect a person's performance. What then is the solution?

Engage members of your team. Discover their strengths, weaknesses, and important details about them. It could've been that you've placed them in a role where they're the weakest. But when you discover their strengths, you'll be able to place them in better roles. As a result, your team will become high-performing and productive.

Follow Due Process

Sometimes, you will have problematic people on your team—no matter how hard you try to make everybody better. When this happens, you should uphold your organization's professional standards and personal composure. Don't get worked up by these people. Instead, keep a record of their actions and take professional measures to address them.

You can contact Human Resources and request their guidance. They'll be happy to help. But if there are no changes, know that you're justified in taking drastic measures. Do this with the knowledge that a great leader can put the organization before his emotions and make difficult decisions for the organization's growth.

Keep Your Composure

As a leader, being frustrated is not the best word to describe how you'll feel when dealing with difficult people. But, since it's often unavoidable, you have to do your best to remain composed and be as professional as possible.

Of course, humans tend to react fast to frustrating times or people. But keeping your cool allows you to escape unnecessary emotional disorders. You can simply get away for a while and calm yourself. Also, you can confide in someone else. This can be a mentor, friend, a team member, or family. This will help you in handling situations more professionally.

Provide a Lift

At times, employees become difficult when their workload is too overwhelming and not motivating. An example is the regular 9 a.m. – 5 p.m. job. An employee who does the same tasks between these hours repeatedly can easily lack the motivation to work effectively.

You can simply arrange shifts or events to relieve the whole team of the everyday stress at work. This should help motivate your team and reshape difficult people for the better.

Lead by Example

Attempting tasks meant for employees isn't a bad idea if you're trying to deal with difficult people. However, some employees actually need training or somebody they can imitate. Showing your team how things should be done can be a good method of relating with them. Also, it creates a thriving atmosphere at work, where employees can approach you.

Studies have shown that people learn better when they see how something's being done rather than just receiving instructions from someone. So boost your team's morale by leading by example.

CHAPTER NINE: CHANGE MANAGEMENT AND OVERCOMING FEAR

"The key to change ... is to let go of fear."

- Rosanne Cash

For your career to thrive, you need to make changes. Failure to make necessary changes will result in you having to depend on expectations and chance instead of plans and statistics. Without making substantial changes, you may not reach your set goals.

Take a look at the world's top organizations and see how they advance. For example, Apple Inc.'s evolution of iPhones from the first model shows that regular changes create massive profits.

In this regard, this chapter will discuss change management, its benefits, and the major hindrance, which is fear.

What Is Change Management?

As a widely used term, change management deals with transitions or changes that occur to people or groups of people. It is geared toward equipping people, teams, and organizations in achieving development. It incorporates techniques influencing the utilization of assets, assigned jobs, or different methods that fundamentally change an organization.

Effective change management involves planning and supporting employees, laying out the essential steps for change, and considering present situations.

Fear

Fear is an inbuilt emotion that can be experienced, to any extent, by people. It acts as an awareness creator as it makes us aware of potential danger, whether real or imagined.

Different situations can trigger fear. Have you ever been in a situation that truly frightened you? Remember the feelings you had. Your heart was beating so hard that you could hear it. You suddenly had energy to move faster. You were alerted to anything around you, and all you wanted to do was escape, right?

Fear can also be linked to feelings of stress and anxiety, which occur when you're uncertain about what the future holds. Sometimes, the outcomes of decisions you've made can cause you to fear. For example, you might be afraid of recruiting new team members, taking risks for the organization, firing difficult employees, and even losing profits. No matter the reason, it's important to subdue fear and use it as a powerful tool in achieving your goals.

Fear arises when you're likely to undergo physical or emotional hurt. Most people consider fear a bad feeling because it mostly arises in negative situations. But the truth is, it actually helps us cope with or escape possible harm.

However, there are fears that you don't count as fear. Those little fears are the ones that wreak havoc. You can be scared to fail, succeed, or change. In some cases, you can be scared of the unknown. Hence, you must employ methods to overcome fear in yourself, your team, and the organization. This will help create an atmosphere of confidence where tasks are being accomplished, trust reigns, and communication is effective.

Here are some tips to help you overcome fear:

Build Your Trust

Albert Einstein once said: *"Every kind of peaceful cooperation among men is primarily based on mutual trust."* This implies that you need to build your trust in people for you not to be in a state of fear. Respect others so that you can gain respect; it results in mutual understanding.

Trust demands that you confide in a person or a group of people. Feel free to relate openly with others through steady physical conversations, online correspondence, and the like. The types of conversations you can try vary from personal to small group chats—to even big team discussions. Through these gatherings, you can learn new techniques for overcoming fear.

Trust makes the team connected. It makes every member of the team rely on one another. When a member makes a decision, their idea is acknowledged, reviewed, and followed. And as a result of this, fear is reduced because all members believe in one another.

Bring Yourself Low in Service

As a leader, you're not expected to be an all-time authoritarian who goes about giving orders, finding the smallest of mistakes, and raining criticism on members when they fall out of line or don't meet the expected objectives. Instead, you should serve so that you can be served too.

According to John Maxwell, *"Anyone can steer a ship, but it takes a leader to chart the course."* So the big question now is:

"What kind of leader are you?" Are you the bossy type—the one that tolerates no shortcomings? Or are you the leader that serves? If you're in a leadership position, overcoming fear in your team begins with you.

A great leader isn't measured by their achievements but by the number of lives they've touched.

Your team already faces fears such as:

- Inabilities to meet deadlines
- Workplace conflicts
- The fear of being fired
- Low organizational productivity
- Fear of disappointments
- Fear of consequences of committing mistakes

A fear-based leadership reduces employees' morale and causes poor performances at work. You don't want them to shiver every time they see you. Instead, it's your job as a leader to build your faith in yourself and your team.

Believe that they're capable and prepared to handle the tasks ahead. Support their decisions in good faith and be interested in their positive output.

This faith over fear mindset allows you to become a servant leader who serves by safeguarding your team and creating a safe place for them to thrive. Leaders that serve are more like caretakers. They ease fears, supply tools, and offer support so the entire team can function optimally.

Lead with Your Strengths and Not Your Weaknesses

Identify and make the best use of your team's strengths. The goal is to overcome fear. Many members are fearful because they haven't found their way around their weaknesses. That's where you come into play.

Of course, this doesn't mean that you have to be the perfect person. You also have your shortcomings. However, you need to work with your strengths and improve your weaknesses. Be an example to others in terms of strengths, and you should be able to delegate tasks that you're not so good at to others who are more specialized than you're. That way, you can show your team that no one is perfect even though strengths are necessary. In addition, it'll help in building confidence in your team.

Building on strengths can be done in so many ways. First, it involves moving from a weak mentality of yourself or your team into a strength-based mentality. However, two things must be considered before you can successfully bank on strengths:

The first thing is the leader's intentions. The leader must be clear about their plans to capitalize on the team's strengths. If you're planning to build on the team's obvious strong points, your members will surely see you as someone who wants the best for them. They'll see you as a source of inspiration and be motivated to move forward.

Secondly, you should consider awareness creation. Many employees can accomplish tasks, but they don't know how. It's now up to you to devise methods to make them see their strengths and put them to good use. The faster you let them know that they can surpass their current abilities and highlight what you've noticed in them, the easier it becomes to have more confident people at work.

Having a strength-oriented mentality isn't as easy as breaking eggs, but the benefits will cover the efforts when introduced into the team. So as a leader, if you're keen on developing greater leadership qualities, consider having a strength-based mindset.

Although telling people about their weaknesses makes them know the specific area they need to develop, it may also lead to fear, bitterness, and hatred. Therefore, I suggest that you train yourself to have a very strong mindset if this happens.

Use Positive Conflict as Opportunities for Growth

Most people think that conflict is a very negative occurrence. This is wrong. Just like the saying *"We'll disagree to agree,"* conflicts are necessary tools in engaging your employees more. It doesn't have to get nasty; a moderate amount of disagreement is okay to ensure diversity of ideas and the organization's overall progress.

Debates and competitions are examples of conflict that can lead to fresh perspectives and growth for a business. In other words, totally eliminating conflicts at work can ultimately hurt a business's growth even though it can temporarily make the work atmosphere peaceful.

Ponder on the words of Laura Stack, a productivity expert, in an Aviation Pros Article: *"When conflict exists, it generally indicates commitment to organizational goals, because the players are trying to come up with the best solution. This, in turn, promotes challenge, heightens individual regard to the issues, and increases effort. This type of conflict is necessary. Without it, an organization will stagnate."*

Conflict arises from creativity. It's a process where the disagreeing parties come together and drop creative ideas to find a solution that meets everyone's needs. This involves understanding each other's needs, imagining the best-case scenario, and determining how to meet that vision. Conflict encourages a deeper investigation of issues.

Ensure That All Members Stay Focused

During a soccer match, every team player needs to focus on the opponent's goal. And to achieve that, each player has to win the ball for their team. Would you be relaxed if your teammate gave up the ball? Of course not!

In the same way, an absence of focus on the organization's objective is a big deal. And if you don't respond to this "big deal" quickly, the whole organization will be affected.

Every member needs to know that others are dedicated to the same goal they are. Team members should have each other's backs in needy times, unclear situations, and scary circumstances. Otherwise, they'll have to operate in constant fear of failure.

You should address this type of fear by considering the mission as well as the morale of the team members.

Encourage Everyone to Hold Others Accountable – But with Respect

As a leader, you need to serve as a role model to your teammates in terms of accountability. Apart from being accountable yourself, hold others accountable—but do it with respect. If you don't show interest in holding others accountable, it'll be hard for the team to do it too. Everyone must engage and take part in the team's accountability.

Being accountable at work involves knowing the organization's goals, taking steps towards them, and being responsible for your own actions. Hence, employee accountability circles around making every employee aware of the organization's goals and charging them with the responsibility of achieving them.

While accountability in teams is highly essential, it also needs to be regulated with the need to give members the liberty to perform their tasks. They shouldn't feel pressured to get things done. Instead, they should be allowed to freely employ methods that will help attain excellence. Encouraging this culture of team accountability helps to build high-performing teams.

Build an Avenue for Positive Responses

You can attend to people's fear by regularly motivating and inspiring them. You should be a confidence booster. It shows people that you recognize their efforts. Consequently, they'll put more effort into developing their skills to achieve more output.

Feedback is a very important part of any leader's skill. While giving feedback is necessary, receiving it is also essential for effective information exchange, performance improvements, and all-around development in the team.

The essence of feedback in the workplace is unrivaled because it helps relay necessary instructions on what to do to make work better while appreciating the efforts of the people in question. It makes you open to your members.

Leadership Strategies for Overcoming Change Management

Knowing what change management entails, it's imperative to know how you can overcome it. Overcoming change management involves having specific plans, knowing how to relate the plans to members, and working toward them.

Below are some strategies you can adopt as a leader in overcoming change management:

Have a Specific Goal and Ensure Effective Communication

Having clear goals is as important as any first step in decision-making. It might look easy, but it's not always a smooth ride. Teams often forget to carefully analyze their goals, consider the team's current situations, and assume that proposed changes have taken place. Then they can accurately gather possible outcomes and might need to cross-check what changes they want to implement.

It's one thing to make the change required and another to conduct a thorough review against the workplace goals and expectations that'll ensure that the change will strategically carry the organization in the right direction. Hence, you need effective communication.

Effectively sharing information makes the steps easier to take and sets up an environment where better comprehension of individuals and team duties occurs. Also, it gives rise to collaboration and helps to bring everyone together in one accord. Sharing a common view of the project and its goals with your team is important in unifying team members' views.

What's the essence of an excellently crafted goal if there is no means to effectively communicate it with others?

Real breakthroughs occur when your team members are consulted and their views are acknowledged and worked upon.

So be ready for the big task. Because the bigger your team, the harder it becomes to communicate effectively.

Strengthen Your Bond with Your Employees

A leader who doesn't share a personal attachment with team members shouldn't expect them to be fully in on the tasks at hand. Getting members on change-oriented plans requires having personal relationships with them, aside from professional grounds. This is because people tend to tune in to change provided they're familiar with the concept, and you carry them along on methods to be adopted.

Even if you consult them and are not convinced of their responses, you're at liberty to make your own decisions. It's better than not giving members a say in change-related matters.

Be Accountable and Transparent

When we discuss accountability, we are talking about taking responsibility for a given situation or action. I'll be shedding more light on some facts about accountability and transparency within workplaces, including effects when not implemented.

In this context, transparency can be a level of openness between leaders and team members. When transparency is introduced into everyday practices at work, there'll be room for mutual respect, trust, smooth exchange of ideas, and stronger interpersonal work relationships.

Members feel appreciated when their leaders are honest. They'll feel more confident about their jobs. And they won't be uncertain about how to follow organizational procedures in achieving the desired changes.

Sharing organizational decisions with members will let them value you as someone worthy of emulation. As a leader, you should act with transparency by ensuring that your team members aren't kept in the dark about matters regarding the organization.

Common Situations When Change Management Is Needed

Change management has been a major organizational function. And there are various areas where change management is deemed fit to ensure that necessary changes are made in organizations. They include:

- Adoption of new technology
- New purchases
- Change in leadership
- Change in rules and regulations guiding the organization
- Unforeseen circumstances like a drastic decline in profits or physical circumstances like a fire outbreak

Benefits of Change Management in the Organization

As you already know, change is a constant phenomenon. It's a process that requires thorough planning and supervision. Therefore, you should know what you stand to gain before venturing into it.

Be specific about the changes you anticipate and how they affect your team or organization before swinging into action. This way, you'll be acting on facts and not on mere assumptions.

Here are some listed benefits of change management:

- It makes it possible for the organization to be swift in customer service.

- It assists leaders, managers, and concerned people in correctly arranging the organization's resources and utilizing them efficiently.

- Change management gives organizations the chance to carefully analyze and cross-check the effects or consequences of planned changes.

- Change management helps organizations initiate the change into their system without disrupting or hindering their regular business affairs.

- It allows leaders to know the minds of their members. As such, it improves members' efficiency and boosts productivity.

- Change management combines time management techniques with planning changes. Thus, minimizing the time needed to enforce a change.

- It manages the risks involved in making changes by reducing the possibilities of negative change impacts to the barest minimum.

- Since it allows leaders to carry their members along, employee performance at workplaces is increased.

- Quality of customer service is improved, and clients are duly served by competent and trained employees.

- It provides a way to foresee, expect, and prepare for future positive or negative occurrences. This way, it helps find ways to tackle those occurrences in advance.

- Change management helps to manage costs involved in potential changes by considering current budgets, current organizational capacity, and future benefits from the change.

- Since organizational change can be called an investment, change management helps make profitable decisions that will lead to increased return on investment (ROI).

- It creates an avenue for leaders, members, and teams to develop.

CONCLUSION

Many people dream of becoming a leader because of the respect that comes with it. However, it's only a fantasy to think that leadership is a bed of roses, especially as a woman. This book has taught you the qualities of good leadership and how important it is to carry your team along and do the needful for the benefit of your organization.

My zeal to write this book arose because I had also been in these situations, and I know the challenges women face in leadership. This book is a complete guide for effective leadership and management in an organization.

We started out with an introductory chapter on the meaning of leadership and leadership styles. In that first chapter, we looked at some traits of successful women in the workplace and businesses, then in Chapter Two, we discussed the misunderstandings people have about managers and leaders. This informative section discussed the differences between managing people and leading people. We also talked about effective leadership, the skills involved, and its importance.

Chapter Three provided ways to deal with difficult people. Here, we dwelled on the fact that difficult people are everywhere, and since we can't avoid them, we should learn to deal with them. We highlighted how to manage difficult people and professional measures that you can take.

In Chapter Four, we presented effective ways to overcome self-doubt and boost your self-confidence as we discussed the techniques for overcoming self-doubt.

Communication was one of the effective techniques to help build a high-performing team. As we discussed this informative concept in Chapter Five, we can see that members of high-performing teams know the essence of their roles. You need to be a good communicator to build a well-performing team.

When putting this book together, I envisaged a guide that will give you everything you need to be confident as a leader, and one effective way is to overcome your fear. In Chapter Six, we talked about building confidence in your team. You have to trust your own gut and lead by example.

In Chapter Seven, we discussed effective communication skills and their importance. As we approach the finishing line, we discussed negotiation in Chapter Eight. This is a good avenue to build your negotiating skills, be more confident, and represent your organization well. Remember to remain as professional as possible when dealing with difficult people in your workplace.

Chapter Nine brought us to an end as we discussed an important aspect that affects organizations and life in general: change. You can't remain in your comfort zone forever. That's why you should strive to overcome fear. Today, I am pleased with myself because my hard work and training as a leader are paying off. I want the same for you; I want you to be a successful leader. Appreciate your new role and embrace a successful life where there's no fear or doubt.

By practicing what you've learned from this book, you add value to yourself. So don't just drop this book after reading it. Always go back and check useful points in the book, then apply them to yourself.

Are you ready to experience leadership at its best? You've been equipped with the right ideas, techniques, and information; you just need to execute them.

I hope you join other women in the world who are already breaking the bias and setting a path for other women to follow.

Best wishes.

Ingram Content Group UK Ltd.
Milton Keynes UK
UKHW022235180523
421997UK00005B/58